The Capricious Cosmos

ALSO BY JOE ROSEN

Symmetry Discovered: Concepts and Applications in Nature and Science
A Symmetry Primer for Scientists
Symmetry in Physics (as editor)
Developments in General Relativity, Astrophysics and Quantum Theory (as editor, with F. I. Cooperstock and L. P. Horwitz)

THE CAPRICIOUS COSMOS

Universe Beyond Law

Joe Rosen

Macmillan Publishing Company
New York

Maxwell Macmillan Canada
Toronto

Maxwell Macmillan International
New York   Oxford   Singapore   Sydney

Copyright © 1991 by Joe Rosen
All rights reserved. No part of this book may be reproduced or transmitted in any form or by any means, electronic or mechanical, including photocopying, recording, or by any information storage and retrieval system, without permission in writing from the Publisher.

Macmillan Publishing Company
866 Third Avenue
New York, NY 10022

Maxwell Macmillan Canada, Inc.
1200 Eglinton Avenue East
Suite 200
Don Mills, Ontario M3C 3N1

Macmillan Publishing Company is part of the Maxwell Communication Group of Companies.

Library of Congress Cataloging-in-Publication Data
Rosen, Joe.
    The capricious cosmos: universe beyond law / Joe Rosen.
    p.   cm.
    Includes bibliographical references and index.
    ISBN 0-02-604931-7
    1. Cosmology—Popular works.   I. Title.
QB982.R67   1992
523.1—dc20   91-19625
                                                              CIP

Macmillan books are available at special discounts for bulk purchases for sales promotions, premiums, fund-raising, or educational use. For details, contact:

    Special Sales Director
    Macmillan Publishing Company
    866 Third Avenue
    New York, NY 10022

10 9 8 7 6 5 4 3 2 1
Printed in the United States of America

Contents

Acknowledgments vii

Introduction ix

1. WHAT IS SCIENCE? 1
   Preliminaries · Definition · Nature · Reproducibility ·
   Predictability · Law · Summary · Bibliography

2. SCIENTIFIC THEORY 19
   Theory · Logical Implication and Truth · Generality
   and Fundamentality · Naturality · Causation ·
   Simplicity and Unification · Beauty · Falsifiability ·
   Archetypical Example · Summary · Bibliography

3. BEYOND SCIENCE 33
   Science and Metaphysics · Transcendence and
   Nontranscendence · Summary · Bibliography

4. UNIQUE UNIVERSE 45
   Capricious Cosmos · Cosmology · Summary ·
   Bibliography

CONTENTS

5. **LAWS OF NATURE** 61
   Realism and Idealism • Reductionism and Holism • Observer and Observed • Quasi-isolated System and Surroundings • Initial State and Law of Evolution • Summary • Bibliography

6. **LAWS OF NATURE (CONTINUED)** 85
   Extended Mach Principle • Whence Order? • Summary • Bibliography

7. **WE AND THE UNIVERSE** 99
   Human Science • Anthropic Principle • Whence Order? (Again) • Space and Time • Summary • Bibliography

8. **REALITY** 117
   Metaphysical Positions • Objective Reality • Perceived Reality • Partially Hidden Reality • Transcendent Reality • Summary • Bibliography

9. **SELF-GENERATING UNIVERSE** 131
   Cosmological Schemes • Baby Universes • Closing the Circle • Many Worlds • Summary • Bibliography

Glossary 143

Index 157

Acknowledgments

I would like to express my warm thanks to those who read early versions of the manuscript, commented, criticized, and offered their advice: Avi Elitzur, Larry Fagg, Jan Fennema, Thomas von Foerster, Nathan Rosen, and two anonymous reviewers. I am especially grateful to Mira Frost, who did all the above and so much more. I would also like to thank the Department of Physics of The Catholic University of America and its chairman, Jim Brennan, for their hospitality during my sabbatical leave, when much of this book was written, and the School of Physics and Astronomy of Tel Aviv University, my "home institution," where this book was conceived and its writing begun. Many thanks are due the Macmillan staff, for their skillful turning of my raw manuscript into this finished book, and especially Natalie Chapman, senior editor, whose helpful suggestions were invaluable.

Introduction

During recent years there have appeared a rather large number of books that present, in a style accessible to the intellectually curious layperson, the modern scientific conception of the fundamental aspects and phenomena of nature and the remarkable and impressive achievements of scientists, more specifically of physicists, in gaining understanding of those aspects and phenomena. Those books are most welcome.

The attainments of science, and especially of physics, since the beginning of the twentieth century have radically changed our conception of nature from what it was previously. They have, in fact, revolutionized our view of the universe, even of reality. While we enjoy many technological benefits accruing from the advances of science, and also suffer from some of the results, it seems to me that it is really in the realm of our intellectual grasp of the world around us and our role in it, in the realm of our philosophy of reality, that the impact of modern science is truly shattering. Yet this intellectual adventure of humankind has been largely confined to the circles of scientists, mostly physicists, and science-oriented philosophers. And that is surely a pity. The general public (which in the final reckoning is paying for it) should be given the opportunity to share this exciting adventure.

Fortunately a number of scientists, among whom are some who are themselves active, even pivotal, in the adventure of modern science, and who possess the all-too-rare talent of being able to explain their field to those willing to make the effort to understand, have written popularizing books in recent years. All praise is due their attempts to share the adventure with those for whom science is not the way of life.

However, I have a serious reservation. It seems to me that the lay reader is too liable to gain the impression that not only is science capable of attaining full understanding of the material universe as a whole, in all its aspects and with all its phenomena, including the role of *Homo sapiens* in it, but that science is actually on the verge of doing so. Some authors believe that themselves. They seriously consider the possibility of a "Theory of Everything," capitalized just like that and acronymed to TOE. One of them even declares that we might "know the mind of God"! Others do not commit themselves, but allow that impression to develop.

Yet the fact of the matter is that science, by its very nature and structure, cannot in principle comprehend the material universe as a whole. Through science we can, and indeed do, gain understanding of various aspects and phenomena of the material universe and discover laws governing them. But as for the whole, as for the material universe in its entirety, it inherently lies beyond science. Within the framework of science the very concept of law governing the universe as a whole is meaningless. By the very character of science an understanding of everything, a "Theory of Everything," even only of everything of material nature, is but a chimera. As far as science is concerned the material universe as a whole is orderless, lawless, and unexplainable, indeed *the capricious cosmos*. Any understanding of the whole can then come only from outside science, from nonscientific modes of comprehension and understanding.

The main theme of this book is that science can go only so far in comprehending nature. We will see just how far science can go. We will see what the domains of science and of metaphysics are, where science leaves off and metaphysics picks up. Yet we will see how science and metaphysics are intimately entangled—how met-

Introduction

aphysical considerations are unavoidable in science and how science influences our metaphysical positions, i.e., what science in fact tells us about reality. And science tells us that it cannot fully comprehend the objective reality underlying the world of our perceptions. That leaves even more room for nonscientific modes of comprehension and understanding.

Another theme of this book is the apparent contradiction, even paradox, of the inherent orderlessness, lawlessness, and unexplainability of the universe *as a whole* through science, on the one hand, and, on the other, the obvious fact that *within* the universe there are order and law that science comprehends and explains. Why are there order and law within the capricious cosmos? This theme will lead us to, among other things, the anthropic principle, which is the use of the existence of human beings as an unconventional explanation of other aspects and phenomena of nature. Following a thorough consideration of its advantages and limitations, we will see the true status of the anthropic principle as a *scientific* explanatory principle.

Yet another theme of this book is the position of humankind in the universe and in science. That connects with the anthropic principle. It also connects with the fact that science is a *human* endeavor and the implications thereof.

Thus this book is a sequel to, a complement for, even an antidote to books of the kind mentioned earlier. I hope you have already read one or more of them. (That would be useful, but if not, never mind.) If you have indeed formed the mistaken impression I am concerned about, I hope to cure that. Otherwise, I hope you will find interest in what I have to say anyway.

My approach is to start from rather basic definitions and concepts connected with science and to develop them through simple logic and common sense to the conclusions I want to convince you of. Since I am a theoretical physicist, my approach and point of view are, not surprisingly, those of a theoretical physicist. Physics, the most fundamental of natural sciences, is my model of a science. Other natural sciences differ from physics and from each other, not only in their subject matter of course, but also in their fundamental conceptions and their methods. Yet it seems to me that there is sufficient basic similarity and overlap in conceptions and methods to justify my using the term science throughout this book to cover natural science in general.

In this book I am presenting my personal view of science, nature, the universe, reality, and the position of humankind in all that—a sort of "The Way Things Are, According to JR," or "How Things Go, According to Joe," or "Rosen's Ruminations on Reality." I have my own opinions and biases and do not hesitate to express them, although I do try to put them in perspective. Some of my views are mainstream, shared by most scientists. Others are more or less unconventional, yet fully justified and completely respectable. Among the latter: Cosmology reaches beyond the domain of science and is basically metaphysics. The origin of the lawful behavior of quasi-isolated systems is with the universe as a whole (the extended Mach principle). The anthropic principle can provide valid, though limited, explanations within science. The self-generating universe scheme provides a conservative, economical conceptual framework for the coming into being of the universe and of its quantum alternatives. For more and for the details you will have to read the book.

The introduction to a book like this is the only opportunity an author has to preempt, outguess, ward off, or disarm his or her unknown future critics. I can already imagine the review: "Rosen's apparent ignorance of the essential distinction between objectivity and intersubjectivity . . ." etc. Well, to the writer of that review I reply in advance that my ignoring the distinction between those two (also called, respectively, strong and weak objectivity) was intentional and fully in line with the objectives of this book. Indeed, I am convinced that for the purpose of the presentation the gain outweighs the loss.

To the philosopher reviewers I say: This book is not for you! You will try to categorize my metaphysical position as watered-down this or half-baked that. You will wonder in print whether I

Introduction

have ever heard of Kant or who knows whom else. Please desist. You are missing the point. I did not write a philosophical treatise. I wrote about science and the very near neighborhood of science. I tried to present certain issues as a scientist, not as a philosopher. If I used terms and concepts a bit loosely by your standards, that was to help the presentation and clarify the issues.

And to those who find my metaphysical position, as declared in this book, to be self-contradictory, please note that I admit that myself. The presentation reflects a real shift in my world view, a shift whose implications I am still trying to fathom. The nontranscendent position presented as mine in chapter 3 was my actual world view. Fairly recently I adopted the world view developed in chapter 8, involving an objective reality shown by science to be necessarily transcendent. The contradiction is discussed in that chapter.

The Capricious Cosmos

1 What Is Science?

Reason's razor rives Nature's reticence
Revealing her awesome order.
O lucid law! O scintillant symmetry!
Your splendor sears the soul.

PRELIMINARIES

Science. *Science.* SCIENCE. What a wonderful, powerful-sounding word! It instantly induces an atmosphere of pure rationality. It rings loudly a symphony of universal knowledge and understanding. It forcefully projects an aura of all-encompassing order and of control. Does the world look a mess? Are things getting out of hand? Never fear; science is here! All will soon be set straight. Order will soon be found and everything will be understood and under control.

Cancer? It will surely be conquered; just give science a few more years and sufficient funds. Hunger? Science already has the problem licked (and it's the politicians who are to blame). Where do we come from? Science tells us authoritatively, and its scenario runs from the big bang, through the materialization of matter and an expanding universe, through the formation of galaxies, stars, and planets, through the first appearance of life on Earth, through evolution, and on to us. Where are we going? Science presently has to admit that it does not yet have the answer. But it certainly will. Oh yes, it will tell us, and then we will know.

Is that what science means to you? Are those the associations science arouses for you? If so, you are in good and numerous company. Very many people, and among them even a large number of scientists, believe that science is the fountainhead of understanding of, if not everything, then at least everything of material nature. However, they are mistaken.

Granted, science is usually not expected to comprehend the nonmaterial aspects of the universe, such as morals, beauty, love, belief, and so on. Most of us realize that and limit our expectations accordingly. And there are aspects of the universe, such as life, consciousness, mind, intelligence, and so on, whose materiality (or nonmateriality) is the subject of debate. Thus it is not clear whether science should be able to comprehend them or not. There are indeed well-founded opinions on both sides of the question, so the issue is very much an open one.

But even in its handling of the material universe, science, by its very nature, has inherent limitations. One such limitation is that it cannot comprehend the material universe as a whole. Science does, indeed, give us understanding of various aspects of the material universe and phenomena within it. But science can go only so far. The material universe *as a whole* is in principle beyond science's grasp.

Nonscientific modes of comprehending and understanding, such as feeling, intuition, and religion, have, with the historical rise of rationalism, become largely relegated to those aspects of the universe that lie outside the domain of validity of science. (I have a mental picture here of throwing scraps to a beggar, of whom one is ashamed.) And I fully concur with that. By my taste, for all those aspects of the universe that science successfully comprehends or is potentially capable of comprehending, nonscientific modes of comprehending should indeed be denied authority as sources of understanding.

As an example, consider the multicolor effect obtained when sunlight passes through a glass prism or through drops of water.

What Is Science?

Science offers an explanation of the phenomenon in terms of the wave nature of light, the dependence of the speed of light in glass and water on the frequency of the transmitted light, and so on. A nonscientific explanation might be that the spectrum is the direct result of God's will to enrich our lives with the beauty of the colors or is a sign of God's promise to humankind not to repeat a devastating flood. And although I am ever anew full of pleasure and wonder at the sight of a spectrum and especially a rainbow, and although the sight of a rainbow might sometimes remind me of a certain biblical narration, I will take the former any time as the only explanation worthy of that title.

Those who prefer explanations involving feeling or belief over scientific explanations in those domains where science is valid are welcome to them. That is their personal business. But the encroachment of nonscientific modes of comprehension on the domain of validity of science as a matter of public policy is another matter altogether. Science, where valid, offers by far the most nearly objective comprehension and understanding available, and is thus the only mode of comprehension suitable for general public recognition. It is a unifying factor for humankind and a firm foundation for a world culture. Nonscientific modes of comprehension, on the other hand, are subjective to the extreme and thus divisive (except in closed homogeneous communities). They are best left to the individual, or to like-feeling and like-believing voluntary groups of individuals, and must not be allowed to become public policy, especially not in the domain of validity of science.

However, as we will see in the following, the material universe *as a whole* lies outside the domain of science. Thus cosmological schemes, schemes offering descriptions of the birth, evolution, and possible death of the universe, as useful as they might be for science—and they are indeed very useful—are beyond the competence of science. They are no more or less valid than analogous descriptions given by religion (one's own) or by myth (the other guy's religion). Since science cannot authoritatively speak about the universe as a whole, we have here an opening for the legitimate entry of nonscientific modes of comprehension into the business of explaining the material universe. Here one's feelings and beliefs can be just as valid as the scientific-appearing descriptions espoused by scientists. They might even be more valid to their holder than the latter, if, for instance, they are more satisfying or esthetically pleasing.

My own background as a physicist makes me partial to cosmological schemes couched in scientific terms. Indeed I have proposed one myself and it is presented later in this book. Even so, I do realize the inherent lack of scientific validity of such schemes and avoid taking them as seriously as I take science. But more about that later on. Anyhow, if someone prefers the biblical description of the coming into being of the universe, for example, or any other description couched in mythic terms, science cannot object. It really can do no better.

In order to understand some of the inherent limitations of science and, in particular, its inability to comprehend even the material universe as a whole, we must first understand just what science is (and is not). That is what the present chapter and the next are about. The "scientific method," that oversimplification taught us in school, involving observation, hypothesis, experiment, and theory, is only part of the picture. The present and next chapters are also only part of the picture, but are the part we need as background knowledge to support our presentation and allow me to make my points. So let us get to work, and since we should know what we are talking about, we start with some definitions.

DEFINITION

Let us consider the following definition of science: *Science is our attempt to understand the reproducible and predictable aspects of nature.*

First look at *our*. That seemingly innocuous qualifier actually carries a heavy load of implication. It tells us that the source of science is within ourselves, that science, although having to do with nature, is actually a human endeavor. Nature, presumably, would go its merry way whether we were around or not or whether we tried to understand it or not. But without our curiosity and urge to understand, *science* would not exist.

Now for *attempt*. That is our admission that we forgo *a priori* any claim for assured success. Thus science, in spite of its amazing successes, just might not be capable of handling everything within its domain.

Next consider *understand*. We take "understand" to mean "be able to explain." That is just fine, but what then is meant by "explain"? Here we use the dictionary definition: "give reasons for." Thus we consider a phenomenon to be understood if we are satisfied that we know the reasons for it.

NATURE

Now consider *nature*. By that term we mean exactly *the material universe with which we can, or can conceivably, interact*. The universe is everything. The material universe is everything having a purely material character. To interact with something is to act upon it and be acted upon by it. That implies the possibility of performing observations and measurements on it and of receiving data from it, which is what we are actually interested in. To be able *conceivably* to interact means that, although we might not be able to interact at present, interaction is not precluded by any principle known to us and is considered attainable through further technological research and development. Thus the material universe with which we can, or can conceivably, interact is everything of purely material character that we can, or can conceivably, observe and measure. That is what we mean by nature.

"But nature is surely more than that!" many would exclaim. "What about truth, beauty, love, etc.? Aren't they part of nature too?" They and such others are certainly part of the universe, but whether they are of purely material character or not is a very open question. So solely for the purpose of our presentation we exclude such concepts as mind, idea, feeling, emotion, and so on, and confine ourselves to the narrow, strictly materialist definition of nature. We do that merely in order to have a convenient, concise term for the subject of our present investigation, which is *the material universe with which we can, or can conceivably, interact.* The universe might very well possess other components too, but if so, they are not of concern to us here.

That leaves *reproducible and predictable aspects* to look into, and this involves some discussion.

REPRODUCIBILITY

Reproducibility means that experiments can be repeated by the same and other investigators, thus giving data of objective, lasting value about the phenomena of nature. Reproducibility makes science a common human endeavor (rather than, say, an incoherent collection of private, incommensurable efforts). It allows investigators to communicate meaningfully and to progress through joint effort. Reproducibility makes science as nearly as possible an objective endeavor of lasting validity. There seems to be no necessity *a priori* that nature be reproducible at all, but the very fact that science is being done proves that nature indeed possesses reproducible aspects.

We do not claim here that nature is reproducible in all its aspects. But any irreproducible aspects it might possess lie outside the domain of science by the very definition of science that forms the basis of our present investigation. Parapsychological phenomena, for example—extrasensory perception (ESP), telepathy, telekinesis, clairvoyance—if, as some claim, they exist, would be such an irreproducible aspect of nature.

For a detailed view of reproducibility let us express things in terms of experiments and their results. Reproducibility is then commonly defined by the statement that the same experiment always gives the same result. But what is the "same" experiment?

What Is Science?

Actually each experiment, and we are including here even each run on the same experimental apparatus, is a unique phenomenon. No two experiments are identical. They must differ at least in their times (the experiment being repeated in the same laboratory) or in their locations (the experiment being duplicated in another laboratory), and might, and in fact always do, differ in other aspects as well, such as in their directions in space. So when we specify "same" experiment and "same" result, we actually mean equivalent in some sense rather than identical. We cannot even begin to think about reproducibility without permitting ourselves to overlook certain differences, which involve time and location as well as various other aspects of experiments.

Consider the difference between two experiments as being expressed by the change that must be imposed on one experiment in order to make it into the other. Such a change might involve a change of time, if the experiments are performed at different times. It might (also) involve a change of location, if they are (also) performed at different locations. If the experimental setups have different directions in space, the change will involve rotation from one direction to the other. If they are in different states of motion, a change of velocity will be involved. We might replace a brass component of the apparatus with an equivalent plastic one. We might bend the apparatus. Or we might measure velocity rather than pressure. And so on.

But not all possible changes are what we associate with reproducibility. Let us list those we do. We certainly want change of time, to allow the experiment to be repeated in the same laboratory, and rotation and change of location, to allow other laboratories to duplicate the experiment. Since almost all laboratories are attached to the Earth, the motion of the Earth—a complicated affair compounded of its daily rotation about its axis, its yearly revolution around the Sun, and, in addition, the Sun's motion in which the Earth, along with the whole solar system, participates—requires rotation and changes of location and velocity both for experiments repeated in the same laboratory and for those duplicated in other laboratories.

Then, to allow the use of different sets of apparatus, we need replacement by other materials, other atoms, other elementary particles, etc. Due to unavoidable limitations on the precision of experiments, we must also include very small changes in the conditions of running the experiments. And then there are additional changes of a more technical nature, which we will not go into here.

Thus the reproducibility-associated changes that might be imposed on experiments include: change of time, change of direction (rotation), change of location, change of velocity, replacement by other materials, and small changes in the conditions.

We now define reproducibility as follows: Consider an experiment and its result and consider the experiment obtained by changing the original one by any reproducibility-associated change (such as change in the time or place the experiment is carried out). Now it might happen that the result of the changed experiment is related to the result of the original experiment in exactly the same way as the changed experiment is related to the original one. That is, it might happen that the result of the changed experiment can be obtained from the original result by exactly the same reproducibility-associated change by which the changed experiment is obtained from the original (such as the same change of time or place). And it might happen that this is true for all reproducibility-associated changes (all changes of time, direction, location, velocity, etc.). If so, we have reproducibility.

The idea is that when any reproducibility-associated change is imposed on a reproducible experiment-result in its entirety—that is, the same change is imposed on both the experiment and its result—the changed result will always be the actual result of running the changed experiment. One could then say that nature is indifferent to reproducibility-associated changes; they are unessential changes. Impose such a change on a reproducible experiment, and you get an experiment that, although it might differ from the original one in location, orientation, time of execution, and so on, is still *essentially* the same experiment in that its result is *essentially* the same result.

As an example, imagine some experiment whose result is a particle appearing at some point in the apparatus some time interval after the switch is turned on. Now imagine repeating the experiment with the same apparatus, in the same direction and state of motion relative to the Earth, etc., but eight and a half hours later and at a location 2.2 kilometers east of the original location. If that particle now appears eight and a half hours later than and 2.2 kilometers east of its previous appearance, we have evidence that the experiment might be reproducible. (Evidence, but not proof of reproducibility. By the rules of logic, whereas a single negative result disproves reproducibility, no finite number of positive results can prove it. A few positive results make us suspect reproducibility; many will convince us; additional positive results will confirm our belief.)

Or, to put things the other way, imagine now some experiment that is known to be reproducible and whose result is an explosion occurring at some point in the apparatus some time interval after the switch is turned on. Then if we repeat the experiment with the same apparatus, in the same direction and state of motion relative to the Earth, etc., but three hours later and at a location 1.6 kilometers west of the original location, we are assured that the explosion will now occur three hours later than and 1.6 kilometers west of its previous occurrence.

PREDICTABILITY

Predictability means that among the natural phenomena investigated, order can be found, from which laws can be formulated, predicting the results of new experiments. Predictability makes science a means both to understand and to exploit nature. Just as for reproducibility, there seems to be no necessity *a priori* that nature be predictable at all, but the very fact that science is being done proves that nature indeed possesses predictable aspects. And just as for reproducibility, neither do we claim here that nature is predictable in all its aspects. But any unpredictable aspects it might possess lie outside the domain of science by the very definition of science that forms the basis of our present investigation. Indeed, parapsychological phenomena, for example, if they exist, would be such an aspect.

In order to view predictability in detail, let us again express things in terms of experiments and their results. Predictability, then, is the possibility of predicting the results of new experiments. Of course that does not often come about through pure inspiration, but is usually attained by performing experiments, studying their results, finding order among the collected data, and formulating laws that fit the data and predict new results.

So imagine we have an experimental setup and run a series of, say, fifty experiments on it. We have experimental inputs $exp_1, \ldots, exp_{50}$ and corresponding experimental results $res_1, \ldots, res_{50}$, respectively. Thus we obtain the fifty data pairs $(exp_1, res_1), \ldots, (exp_{50}, res_{50})$. We then study these data, apply experience, insight, and intuition, perhaps plot them in various ways, and, perhaps with a bit of luck, we might discover order among them. Suppose we find that all the data pairs obey a certain relation such that all the results are related to their respective inputs in the same way. That relation is then a candidate for a law predicting the result *res* for *any* experimental input *exp*: simply apply the relation that works so well for the data pairs already obtained to any input *exp* and thus find the predicted result *res*. Imagine further that this is indeed the correct law. Then additional experiments will confirm it, and we will find that data pairs (exp_{51}, res_{51}), (exp_{52}, res_{52}), and so on, also obey the same relation, as predicted. Predictability is the existence of such relations for experiments and their results.

For an example of that, consider the experimental setup of a given sphere rolling down a fixed inclined plane. The experimental procedure consists of releasing the sphere from rest, letting it roll for any time interval, t, and noting the distance, d, the sphere rolls during that time interval. Here t and d are playing the roles of *exp* and *res*, respectively. Suppose now we perform ten experiments, giving the data pairs $(t_1,d_1), \ldots ,(t_{10},d_{10})$. Or, to be more specific, say we obtain the following ten numerical data pairs, where the time interval, t, is measured in seconds (s) and the distance, d, is in meters (m):

t(s)	0.5	1.0	1.5	2.0	2.5	3.0	3.5	4.0	4.5	5.0
d(m)	0.025	0.100	0.225	0.400	0.625	0.900	1.225	1.600	2.025	2.500

We study these data and plot them in various ways. (The example is idealized, since in the real world there are always experimental errors that must be dealt with. But for the sake of not unduly complicating our presentation we ignore such difficulties.) Most of the plots show nothing especially interesting. But lo and behold! In the plot of distance, d, against *square* of time interval, t^2, it looks as if all ten points tend to fall on a straight line:

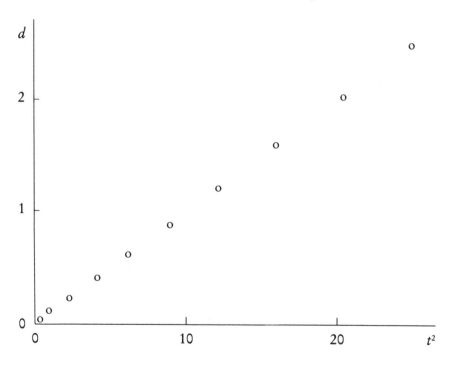

Now if you find yourself getting bogged down in the following mathematical details, do not be overly concerned. Just skip them. I think the point will be apparent in spite of that. The fact we just discovered, that the ten data points all tend to fall on a straight line in the plot of d against t^2, suggests the relation that the distance traveled from rest is proportional to the square of the time interval: $d_1 = bt_1^2, \ldots, d_{10} = bt_{10}^2$, where b is the coefficient of proportionality in units of meters per second per second (m/s²). For our numerical data we would find that $b = 0.10$ m/s², and this relation would be:

t	0.5	1.0	1.5	2.0	2.5	3.0	3.5	4.0	4.5	5.0
t^2	0.25	1.0	2.25	4.0	6.25	9.0	12.25	16.0	20.25	25.0
$0.10t^2$	0.025	0.10	0.225	0.40	0.625	0.90	1.225	1.60	2.025	2.50
$= d$	0.025	0.100	0.225	0.400	0.625	0.900	1.225	1.600	2.025	2.500

That suggests the law $d = bt^2$ predicting the distance, d, for *any* time interval, t. For the numerical example the suggested law is $d = 0.10t^2$. As it happens, this hypothesis is correct, and all additional experiments confirm it: $d_{11} = bt_{11}^2$, $d_{12} = bt_{12}^2$, and so on. Or, we might run the experiment for the additional time intervals $t = 5.5$ s, $t = 6.0$ s, etc. By substituting those values of t in the relation $d = 0.10t^2$ we calculate the respective predicted distances the sphere will roll for the additional time intervals: $d = 0.10 \times 5.5^2 = 3.025$ m, $d = 0.10 \times 6.0^2 = 3.60$ m, and so on. And indeed, when the experiment is run, those are the distances actually measured: $d = 3.025$ m for $t = 5.5$ s, $d = 3.600$ m for $t = 6.0$ s, and so on. The new data are found to obey the suggested law. Thus the relation of distance to time interval is a predictable aspect of the setup.

It should now be clear what we mean by *the reproducible and predictable aspects of nature*. They are those aspects of nature that are both objective (all investigators agree on them) and orderly (they exhibit sufficient regularity to allow prediction). But note the following. Since we do not include *the reproducible and predictable aspects* in our definition of nature as *the material universe with which we can, or can conceivably, interact,* it follows that we are leaving the door open to the possible appearance of irreproducible and/or unpredictable phenomena in nature as our field of interest. Such phenomena could be very interesting and we might very well, and with good reason, attempt to understand them. But then we would not be doing science; we would be involved in nonscientific modes of comprehension and explanation, and that would belong to another story altogether.

LAW

At this point we should be in a position to appreciate the definition of science that forms the basis of our present investigation: *Science is our attempt to understand the reproducible and predictable aspects of nature.* In our effort to understand we first search for order among the reproducible phenomena of nature and attempt to formulate laws that fit the collected data and predict new results. Such laws of nature are expressions of order, of simplicity. They condense all existing data, as well as any amount of potential data, into compact expressions. Thus, they are abstractions from the sets of data from which they are derived, are unifying, descriptive devices for their relevant classes of natural phenomena.

In the above example of the rolling sphere the law $d = bt^2$ is an abstraction from the data pairs (d_1,t_1), (d_2,t_2), ..., a simplification of them. It expresses an order existing among the different runs of the experiment. It offers a description and a unification of the rolls of the sphere down the plane.

As an archetypical example of law of nature derived from experimental data, let us consider the case of Kepler and his three laws of planetary motion. Johannes Kepler (1571–1630) pondered the astronomical data available to him on the five planets that were known at that time—Mercury, Venus, Mars, Jupiter, and Saturn—and found the following order. If the Earth is considered as one of the planets and if the motions of those planets, the six innermost planets of the solar system, are considered from the point of view of a hypothetical observer standing on the Sun (the heliocentric point of view), then those motions are found to possess three properties. For our present discussion it is not necessary to understand those properties in detail, and if you are unfamiliar with the concepts involved, do not worry. The properties are:

(1) *The path each planet traverses in space, its orbit, lies wholly in a fixed plane and has the form of an ellipse, of which the Sun is located at a focus.* (Actually, the ellipses of the six planets are very close to being circles, with the Sun located at their common center.)

(2) *As each planet moves along its elliptical orbit, the (imaginary) line connecting it with the Sun sweeps out equal areas during equal time intervals.* Those areas have the shapes of slices from an elliptical pie. (Thus, from geometric considerations, a planet moves faster when it is closer to the Sun and more slowly when it is farther from the Sun. Actually, since the orbits of the six planets are nearly circular, that property simply means that each planet moves with nearly constant speed.)

(3) *The ratio of the squares of the orbital periods of any two planets equals the ratio of the cubes of their respective orbital major axes.* The orbital period of a planet is the time it takes to complete one revolution around the Sun, the "year" of that planet. The orbital major axis of a planet is the major axis of the ellipse formed by its orbit, which is the distance between the two ends of the ellipse, one end being the point of nearest approach to the Sun and the other the point of farthest departure from the Sun. (Since the orbits of the six planets are nearly circular, the major axis is nearly the diameter of the orbit, so without much error one may read "orbital diameters" for "orbital major axes." And since those

lengths are appearing in a ratio so that a power of two cancels, one may also read "orbital radii.") If we denote the orbital period of any one of the planets by T_1 and that of any other by T_2 and denote their respective orbital major axes (or diameters or radii for the six planets) by a_1 and a_2, then that property is expressed by the formula

$$T_1^2/T_2^2 = a_1^3/a_2^3,$$

or equivalently

$$(T_1/T_2)^2 = (a_1/a_2)^3.$$

As I mentioned just before presenting the three properties, it is not necessary to understand the properties in detail. The most important point is that Kepler formulated properties of the planets' motions. It is of secondary importance that those properties concern (1) the form of the planetary orbits, (2) the speed of each planet as it moves along its orbit, and (3) the relation between the time it takes each planet to complete one revolution around the Sun (the "year" of each planet) and the size of its orbit.

Those properties express an order among the astronomical data. They offer a description and a unification of them. The motions of the six planets are not just any motions, but are related by their possession of the three properties.

Those properties are called Kepler's laws of planetary motion. They are laws of nature in that they correctly predicted the relevant properties of the motions of the additional planets that were discovered in the solar system: Uranus, Neptune, and Pluto. They are also laws in that they are found to be valid for any system of astronomical bodies revolving around a massive central body, such as the moons of Jupiter revolving around Jupiter.

SUMMARY

Science is our attempt to understand the reproducible and predictable aspects of nature, where nature is taken to mean the material universe with which we can, or can conceivably, interact. The operation of science involves first searching for order among the reproducible phenomena of nature and then attempting to formulate laws that fit the collected data and predict new results.

BIBLIOGRAPHY

For an idea of what physics, my model of a natural science, is about, see: chapter 1 of **R. K. Adair, *The Great Design: Particles, Fields, and Creation*** (Oxford: Oxford University Press, 1987); and chapter 11 of **J. S. Trefil, *Reading the Mind of God: In Search of the Principle of Universality*** (New York: Charles Scribner's Sons, 1989), which has to do with physics and "the arrogance of theoretical physicists."

For more reading about the nature of science, see: **R. Morris, *Dismantling the Universe: The Nature of Scientific Discovery*** (New York: Simon and Schuster, 1983); chapter 1 of **H. Fritzsch, *The Creation of Matter: The Universe from Beginning to End*** (New York: Basic Books, 1984); chapter 2 of **J. C. Polkinghorne, *One World: The Interaction of Science and Theology*** (Princeton, N.J.: Princeton University Press, 1986); book 1 of **I. Prigogine and I. Stengers, *Order Out of Chaos: Man's New Dialogue with Nature*** (New York: Bantam Books, 1984); and chapter 3 of Trefil, above.

Although for the purpose of the present book it is not necessary to understand Kepler's laws, or even to be familiar with them, the interested reader is directed to any elementary college textbook on mechanics. For a historical perspective of those laws, see chapter 10 of **D. Park, *The How and the Why*** (Princeton, N.J.: Princeton University Press, 1988).

For the significance of the laws of nature and for symmetry as an expression of simplicity of those laws, see: chapters 1, 2, and 4 of R. P. Feynman, *The Character of Physical Law* (Cambridge, Mass.: MIT Press, 1965); as well as chapter 19 of Park, above. More on laws of nature can be found in chapters 5 and 6 of J. D. Barrow, *The World Within the World* (Oxford: Oxford University Press, 1988); and part 3 of H. R. Pagels, *The Cosmic Code: Quantum Physics as the Language of Nature* (New York: Simon and Schuster, 1982).

2 Scientific Theory

Does beauty underlie it all?
Is that the primal principle?
Does beauty's truth uphold the world
That manifests truth's beauty?

We continue from the preceding chapter in our discussion of what science is about.

THEORY

Laws of nature are clearly worthwhile achievements in their own right. Besides their potentially useful predictive power, they offer a unifying description of natural phenomena. But we are not satisfied with that. We want to *explain* laws of nature, to know the reasons for them; we want to *understand* the reproducible and predictable aspects of nature, not just describe and predict their phenomena. That is science. Scientists' term for an explanation of a law of nature is *theory*.

There are a number of criteria by which theories are judged for their acceptance or rejection in science. Those criteria are not inherent to nature itself, are not imposed *on us* by nature in any way, but are imposed *by us* in our search for understanding. Those criteria, however, are but rationalization. What really and basically determines a theory's acceptability is simply its giving us the *feeling* that something is indeed being explained, so that it satisfies our curiosity about the reasons for whatever law we are trying to understand. The detailed criteria are our attempt to rationalize that feeling. Yet it is very difficult, probably impossible, to meaningfully communicate such feelings, and we have no choice but to discuss the criteria.

Feeling? Satisfy? What is going on here? Are we discussing science or not? Aren't those terms associated with nonscientific modes of comprehension? Yes, they are. And here they are connected with the irrational aspect of human behavior. Since, as we saw in the section *Definition* of the previous chapter, science is a human endeavor, all aspects of human behavior can manifest themselves in the doing of science, even irrationality. We try to rationalize our irrationality, of course. But it is unavoidably there, just as in any other human activity. We will see even more of it further on. Yet in spite of that, the scientific mode of comprehension is still the most nearly objective mode of comprehension around, and we will have to make do with it. Actually, I think we make do very well indeed. Nevertheless, an open-eyed realization of the irrational aspects of science as a human activity cannot but benefit our attempt to clarify the role of science in our understanding of the world around us.

LOGICAL IMPLICATION AND TRUTH

The most important property of an acceptable theory, really the *sine qua non*, is that whatever is explaining must logically imply that which is being explained. A "theory" that does not satisfy that criterion cannot be considered to be an explanation at all, cannot be seen as giving reasons for anything. That should be obvious. Still, an example of such a "theory" could be: Planets are similar to apples. There are varieties of apples that are red. The word "red" is spelled with three letters. Kepler's laws of planetary motion are also three in number. Thus Kepler's laws must hold.

There is more here than meets the eye, however. If a theory logically implies more than it was originally intended to explain, then it is predicting new laws. Those new laws must be valid laws of nature, otherwise the theory is false. Thus, related to the property of logical implication and part of the *sine qua non*, is the other most important property of an acceptable theory, the property of being true in the sense that nothing implied by a theory should contradict experimental findings. That property is also related to the property of falsifiability, to be discussed later on.

As an example, assume we have a theory to explain Kepler's laws of planetary motion and assume that theory also predicts laws of motion for the comets, although cometary motion was not in mind when the theory was devised. Then if the actual motion of the comets does not obey the laws predicted by the theory, the theory is deemed false and thus unacceptable, in spite of its apparent success in explaining planetary motion.

GENERALITY AND FUNDAMENTALITY

Two more properties of an acceptable theory, not *sine qua non* but still very important indeed, are that what is explaining should be more general than what is being explained and should also be more fundamental than the latter. Generality is usually easy enough to discern: the more general the category, the larger the number of natural phenomena it encompasses. Generality endows theories with the property of explaining more than they were originally intended to explain, thus predicting new laws. That allows the theory to be tested by comparing those predicted laws with experiment. The property of a theory that it can be tested in that way is called falsifiability, and we discuss it in more detail further on.

Fundamentality is a less simple matter than generality, since it is dependent on one's scientific world view. Scientists working within the commonly accepted scientific world view usually agree on questions of fundamentality, but even then disagreements can arise. (Note the subjectivity, thus irrationality, here.)

Example: A theory of Kepler's laws based on the motion of the Earth would not be acceptable, since the motion of a single planet is less general than the motions of all the planets. On the other hand, a theory of Kepler's laws based on universal laws of motion for all bodies could be acceptable, because laws of motion for all bodies are more general than laws of motion valid only for planets. Such a theory might then predict laws of motion for, say, comets, and those laws could then be compared with the actual motions of comets.

Another example: A theory of genetics based on some macroscopic property of the organism, such as body weight, would not be acceptable, because heredity is generally considered to be more fundamental than, and to determine to a certain extent, the macroscopic and microscopic properties of organisms. However, since biochemistry is commonly considered to underlie biological phenomena, a molecular theory of genetics, such as the currently very successful one involving DNA, RNA, etc., could be acceptable.

The following is an example of disagreement over fundamentality: There is a class of theories that attempt to explain aspects of the universe as a whole by the existence of human beings. Those theories are unacceptable to many scientists, as the existence of human beings is generally considered to be much less fundamental than anything having to do with the universe as a whole. Proponents of those theories claim, however, that in a certain sense our existence can be conceived as being more fundamental than the universe as a whole. That approach is called the anthropic principle and is discussed in detail in chapter 7.

Scientific Theory

NATURALITY

Another important property of an acceptable theory is that, just as what is being explained is an aspect of nature (where nature, let us recall, has the very specific and narrow meaning we assigned it in the section *Nature* of the previous chapter), so should what is explaining possess a natural character. In other words, a theory should explain one aspect of nature by another and should not look outside nature for its explanations. For example, "the apple falls by the will of God" is not accepted as a theory among scientists, since godhood, by its very definition, lies outside nature. And thus also does creationism fail as a scientific theory. But "our fate is in the stars," if it were true, could be an acceptable theory.

The criterion of naturality is not always strictly adhered to, however, and there are controversies among scientists about just how loose a theory can be in this regard and still be acceptable as a scientific theory. The problem has to do with theories concerning aspects of nature that are generally considered to be among the most fundamental of all, such as space and time, the properties of the elementary particles, and the evolution of the universe. Very simply, in order to explain a most fundamental aspect of nature by something even more fundamental, one is forced to go beyond nature.

Thus one finds modern theories that are based on inherently undetectable extra spatial dimensions, that involve other universes with which we cannot conceivably interact, that consider the situation prevailing "before" the coming into being of the universe and thus "outside" space and time themselves, for example. There are, however, scientists who consider such theories unacceptable, who feel that no explaining is really being done by them. Whether those ideas are accepted as explanations or not, they certainly do offer unifying frameworks, tying together diverse aspects of nature such as the properties of the elementary particles and the evolution and large-scale properties of the universe. So we see that in such a pinch some scientists tend to prefer to stick with fundamentality even at the expense of naturality.

CAUSATION

Another property of an acceptable theory is that what is explaining should be perceived not merely as logically implying what is being explained but as actually causing it. That means that some sort of causal linkage, some "mechanism," should be perceived as joining what is explaining, as the cause, with what is being explained, as the effect. Note that it is the *perception* of such a causal linkage that enhances the acceptability of a theory. Whether there "actually exists" a causal mechanism or not can depend on one's point of view, even on one's scientific world view, and so is not an objective property of nature. (Again note the subjectivity here.)

For example, a theory explaining Kepler's laws might consist of a statement of general laws of motion valid for all bodies and not just for planets. Kepler's laws, describing the motion of the solar system as a special case, would be derivable by mathematical means from those general laws. Thus, although the theory would have the property of logical (mathematical) implication, it would most likely not arouse the perception of causation, so that some might find such a theory not completely satisfying. On the other hand, a theory of Kepler's laws might be stated, from a somewhat different point of view, as general laws of motion together with the existence of the Sun and the planets' attraction to the Sun. Such a theory would most likely arouse the perception of the Sun's causing the planets' special motion through its pull on them. As a matter of fact, the standard theory of Kepler's laws is just of that kind, as we will see soon.

SIMPLICITY AND UNIFICATION

Two additional properties of an acceptable theory are that what is explaining should be simpler than what is being explained and should also be more unifying than the latter. And the simpler and more unifying it is, the more acceptable the theory. Although some criteria for simplicity can be stated (but we will not), a generally satisfactory objectification of that concept has not been achieved. Simplicity is largely a matter of subjective perception. (Yes, subjectivity again.) Simplicity depends very much on one's taste and world view, even on one's education, although there does seem to exist a considerable degree of consensus about it among scientists working in the same field. In any case, scientists prefer "simple" theories to "complicated" ones, however simplicity is judged.

For example, Albert Einstein's (1879–1955) general theory of relativity is one of a number of proposed theories of gravitation, the universal force of attraction between all pairs of bodies in the universe, including the force holding us firmly down to Earth so we do not float off into space. Although all those theories might seem overwhelmingly complicated, even to scientists who are not theoretical physicists, among the initiated it is Einstein's theory that is generally perceived as the simplest and thus as the preferred one.

Unification is easier than simplicity to pinpoint: as a general rule, the more numerous the different concepts a theory involves, the more unifying it is. Thus a theory should tie together and interrelate more aspects of nature than are tied together and interrelated by what is being explained by the theory. For example, the concepts involved in Kepler's laws are purely kinematic, they are solely concepts of motion: position, orbit, area, time, speed, etc. Any explanation involving only those same concepts would not be more unifying than Kepler's laws themselves. But a theory involving kinematic concepts along with additional ones, such as force and mass, would be more unifying. Unification can generally be expected to go hand in hand with generality.

BEAUTY

We now come to a subject that might seem absolutely amazing, and that is the subject of beauty in theories. To many, science has the image of a cold, rational endeavor, to which considerations such as esthetics are perfect strangers. Yet, although rationality is indeed the major ingredient of science, esthetic considerations are far from foreign to it. (We were warned earlier in the present section to expect irrationality in science!) Scientists are constantly heard referring to "beautiful ideas," "beautiful experiments," "beautiful laws," and "beautiful theories." And a scientist will always prefer a beautiful theory to an ugly one, other things being equal, and even often at the expense of some other desirable property of acceptable theories. In fact, many scientists will admit that the pleasure they derive from their profession has a large esthetic component, and for some (including myself) the esthetic consideration is predominant.

What then is a beautiful theory? Well, here we go again. Just as an acceptable theory is one that gives the feeling that something is being explained, so is a beautiful theory a theory that arouses the feeling of beauty. Beauty is not an objective property of theories, but is wholly in the eyes of the beholder. So the most we can do is to rationalize again and try to point out those properties of theories that seem to contribute to their perceived beauty. It appears to me that the main contribution to the beauty of a theory comes from its simplicity, its unification, and its generality. A theory deemed very beautiful by scientists working in the relevant field is invariably very simple, greatly unifying, and of broad generality. Scientists' preference for beautiful theories is truly astounding, when we consider that beauty is subjective and that there *seems* to be nothing in nature that requires beauty. That preference finds expression in scientists' irrational and objectively unfounded conviction that nature *must* be understandable in terms of beautiful theories. The alternative is simply unacceptable.

Scientific Theory

Even more astounding is the fact that successful theories do indeed tend to be beautiful! Or perhaps that is not so astounding after all, considering that theories are part of science and science is a human endeavor. But more about that in chapter 7. In that connection we should relate the story of Paul Adrien Maurice Dirac (1902–84), who developed a very beautiful theory of the electron, a type of elementary particle that is one of the constituents of atoms and is, by its relatively free motion in certain metals, the carrier of electric current in cables and wires. Dirac's theory predicted the existence of another type of elementary particle, very different from the electron in at least one respect (having the opposite sign of electric charge), yet identical to the electron in other respects (such as mass and magnitude of electric charge).

At that time the only other types of elementary particle known were the proton and the neutron, both being constituents of the nuclei of atoms, and neither fitted the specifications of Dirac's predicted particle. Thus the theory was considered false. But Dirac did not give up his theory, and the eventual discovery of the positron proved him right. The moral, according to Dirac, is that "it is more important to have beauty in one's equations than to have them fit experiment."

One who did not follow Dirac's policy and lived to regret it was Erwin Schrödinger (1887–1961). Schrödinger devised a beautiful theory to explain atomic phenomena, but when he applied it to the hydrogen atom, the simplest atomic system, he obtained results that were in disagreement with experiment. Then he noticed that a rough approximation to his equation gave results that agreed with experimental observations. So he published his approximate theory, a much less beautiful one than the original. Due to his delay the original theory was published by others and credited to them. What happened was that the original, beautiful theory was just not appropriate to the electron in the hydrogen atom, but was suitable to types of elementary particles that had not yet been experimentally discovered. The approximate, uglier theory was insensitive to the differences between the types of particles and was fairly accurate when applied to the hydrogen atom.

FALSIFIABILITY

Now we come to the subject of falsifiability, to which we referred previously. Falsifiability amounts to testability, the property that a theory can be tested against as yet unknown natural phenomena to determine whether it is true or false. That property is a technical demand, not directly related to a theory's giving the feeling that something is being explained, but generally required of an acceptable theory nevertheless. In order to be falsifiable a theory must predict something in addition to what it was originally intended to explain (and which it presumably does explain; otherwise it would not be a candidate for an acceptable theory). That prediction is tested against experimental results that were not taken into account when the theory was devised (either because they had not yet been obtained or because they were not known to the deviser of the theory). To be able to produce predictions, a theory must fulfill the criterion of generality, as we saw above. Then what is explaining, being more general than what is being explained, can explain more than it was originally intended to and thus make testable predictions.

A falsifiable theory is in constant danger of being invalidated by even a single new experimental result. An unfalsifiable theory, on the other hand, is, by its very unfalsifiability, immune to invalidation. As an example of an unfalsifiable theory, imagine that a theory of the electron is found that explains all the known properties of the electron, but, unlike Dirac's theory, predicts absolutely nothing in addition to that. Such a theory could not be tested. Even if new properties of the electron are eventually discovered, the theory will not be invalidated thereby. It will still be valid, since it correctly explains what are then only part of the electron's properties, while it has nothing at all to say about the other properties, especially it does not predict their nonexistence. As a putative theory of the electron it will then be considered incomplete, but it will not lose its validity. A falsifiable theory is discussed in the following example.

ARCHETYPICAL EXAMPLE

An archetypical example of a theory explaining a law of nature is the explanation of Kepler's laws of planetary motion by Newton's laws of motion and gravitation. Isaac Newton (1642–1727) pondered Kepler's laws of planetary motion, presented in the section *Law* in the previous chapter, and succeeded in formulating a number of laws of his own in order to explain them. Newton's laws were not based directly on experimental data, as were Kepler's, but were based on Kepler's laws themselves as data. In fact, Newton's laws are somewhat counterintuitive, and their direct experimental verification had to wait for future technological developments and refinements.

Newton proposed three universal laws of motion, whose modern formulation is more or less as follows:

(1) *In the absence of forces acting on it or when such forces cancel each other, a body will remain at rest or continue to move uniformly in a straight line.*

(2) *A force acting on a body will cause the body to undergo acceleration whose direction is that of the force and whose magnitude is proportional to that of the force divided by the body's mass.* Acceleration is change of velocity—i.e., change of speed and/or direction of motion—per unit time. Mass measures the amount of matter in a body; it is related to, though essentially different from, the body's weight.

(3) *For every force acting on it a body will react upon the force's source with a force of opposite direction and equal magnitude along the same line of action.*

To those Newton added the law of universal gravitation: *Every pair of bodies undergoes mutual attraction, with the force acting on each body proportional to the product of the bodies' masses and inversely proportional to the square of their separation.*

For our present purpose it is not necessary to understand the laws in detail and please do not worry if the concepts are unfamiliar. The important point is that Newton formulated universal laws of motion and a law of universal gravitation, valid for all bodies in the universe.

Those four laws form Newton's theory to explain Kepler's laws of planetary motion (as well as a vast realm of other mechanical phenomena). How are they a theory? First of all, they do logically imply the latter. With the appropriate mathematical tools it is easily shown that the motions of bodies around a massive body will, under certain conditions fulfilled by the planets in their motion around the Sun, by Jupiter's moons in their motion around the planet Jupiter, and by other such systems, obey Kepler's laws. Then, Newton's laws are certainly more general than Kepler's in that they hold for all bodies (hence the qualifier "universal") and not just for planetary systems. As a result of that generality Newton's laws explain more than they were originally intended to; they explain a wealth of mechanical phenomena, earthbound as well as astronomical. And they are more unifying, since they show order among broader classes of phenomena than do Kepler's laws: whereas the latter involve solely concepts of motion, Newton's laws involve also the concepts of force and mass. And they are also more unifying in that, for example, the motion of comets is shown by Newton's laws to be akin to the motion of the planets, while Kepler's laws ignore cometary motion.

Whether Newton's laws are more fundamental and simpler than Kepler's is a matter of one's world view and taste concerning fundamentality and simplicity, but scientists generally consider them as such. We will not go into details here, but let us just note in connection with fundamentality that, whereas Kepler's laws merely describe motion, Newton's laws have to do with the causes of motion, i.e., with forces (and their absence). The character of Newton's laws is just as natural as that of Kepler's, and so we have one aspect of nature being explained by another. As they are stated, Newton's laws might not immediately arouse the perception of causation. However, they imply that the motion of every body is determined by all the other bodies through its being gravitationally attracted to them (and possibly also affected by them by means of additional kinds of force). Thus all other bodies are perceived as the cause of a body's motion by the mechanism of gravitational attraction (and possibly other forces). Newton's laws do indeed give scientists the feeling that Kepler's laws are being explained; they satisfy scientists' curiosity about the reasons for Kepler's laws to hold. In addition to all that, Newton's laws possess the property of falsifiability, since they predict so much more than they were originally intended to explain.

SUMMARY

After finding laws of nature, we then we try to understand, i.e., explain, those laws by means of theories. For a theory to be acceptable, whatever is explaining must logically imply that which is being explained. The former should be every bit as much an aspect of nature as the latter. In addition, the former should be more general, more fundamental, more unifying, and simpler than the latter and should be perceived as causing the latter. Beautiful theories are preferred. Theories should be falsifiable.

BIBLIOGRAPHY

Although for the purpose of the present book it is not necessary to understand Newton's laws, or even to be familiar with them, the interested reader is directed to any elementary college textbook on mechanics. For a historical perspective of those laws, see chapters 10 and 11 of D. Park, *The How and the Why* (Princeton, N.J.: Princeton University Press, 1988).

For scientific theory, see chapters 2–4 of F. Rohrlich, *From Paradox to Reality: Our New Concepts of the Physical World* (Cambridge: Cambridge University Press, 1987).

For beauty in science, specifically in physics, see A. Zee, *Fearful Symmetry: The Search for Beauty in Modern Physics* (New York: Macmillan, 1986), especially part 1. For a more introductory presentation of symmetry, see J. Rosen, *Symmetry Discovered: Concepts and Applications in Nature and Science* (Cambridge: Cambridge University Press, 1975).

3 Beyond Science

To soar above Nature's narrow confines;
To leap beyond matter's imprisoning pale;
To free invention's rein and let
The mind weave wondrous worlds.

SCIENCE AND METAPHYSICS

The definition of *science* presented in the section *Definition* of chapter 1 and forming the basis of our present investigation is *our attempt to understand the reproducible and predictable aspects of nature*. The definition of *nature* is strictly limited for the purpose of our investigation to *the material universe with which we can, or can conceivably, interact*, which is what our investigation is concerned with. Yet in the same chapter we mentioned going beyond nature and we referred to world view, taste, fundamentality, simplicity, perceived causation, and beauty, all outside the strict limits of nature, thus beyond the domain of science. That brings us to *metaphysics*.

Metaphysics is a branch of philosophy dealing with being and reality. In this book we use the term metaphysics in its narrower sense of the philosophical framework in which science operates. In that sense metaphysics is concerned with what lies around, below, above, before, and beyond science. For instance, while science involves observation of nature, metaphysics might consider the significance of the observer-observed dichotomy (since observers, after all, are also part of nature). Or, while science searches for and finds order and laws of nature, metaphysics might consider what constitutes evincive evidence, convincing confirmation, and persuasive proof and why there are order and laws of nature at all. Or again, while science is our attempt to understand the reproducible and predictable aspects of the material universe with which we can, or can conceivably, interact, metaphysics might consider modes of existence other than material existence, aspects of the universe with which we cannot, even conceivably, interact, or irreproducible or unpredictable aspects of the universe. Thus concepts such as world view, taste, fundamentality, simplicity, perceived causation, and beauty, which lie beyond the domain of science, belong to the domain of metaphysics.

Considerations of science and considerations of metaphysics—although each kind of consideration is perfectly respectable in its own domain—must not be confused with each other. It behooves scientists to stick to scientific considerations, avoiding metaphysical considerations, for as wide a range of phenomena as possible, but not exceeding the domain of science. And when we choose to do metaphysics, we must be sure the subject of our investigation lies beyond the domain of science. The domain of science is delineated very clearly, by the definition of science, as the reproducible and predictable aspects of the material universe with which we can, or can conceivably, interact. Any phenomenon, concept, etc., that does not toe that line lies beyond the domain of science. Thus science is not concerned with phenomena or concepts that are not aspects of the material universe (such as God, and possibly feeling), or are aspects of the material universe but we cannot, or cannot conceivably, interact with them (such as other island universes, described in the section *Capricious Cosmos* of the next chapter), or are unpredictable (such as ESP), or are irreproducible (such as ESP again). And if such a phenomenon, concept, etc., is related to the philosophic framework in which we do science, it is then grist for the mill of metaphysics. So theories, for example, are scientific stuff, while their possible beauty is a metaphysical matter.

A world view is one's attitude toward and interpretation of reality; it is the conceptual framework by which one organizes one's perceptions. World views lie well within the domain of metaphysics, according to our usage, since they can strongly influence the way science is done. As was mentioned in the previous chapter, issues of taste, fundamentality, simplicity, perceived causation, and beauty in science are subject to one's world view and may well be decided differently by different people.

In science there are rather strict criteria for truth. Whatever one might think or whatever controversy might arise concerning nature, in the final analysis experiments are performed, observations are made, and it is nature itself that is the supreme arbiter. I would be remiss here if I did not warn that the matter is not as clear cut as it might appear from the previous two sentences. There are questions and controversies, mostly of a metaphysical character, about just what scientific truth means (if anything), how it might be attained through science (if it is attainable at all), and so on. But none of that is important for our discussion, nor will it detract from the points I will make. On the whole, scientists have no problems with such issues.

However, the situation is not the same in metaphysics, including world views. Metaphysical considerations are unimpeded by any burden of need to be "true," although they might be expected not to contradict scientific truths and to be logically self-consistent, i.e., not to contradict themselves. At least I would expect that. But even those expectations have no substantial foundation, and one can find metaphysical positions that do contradict the truth as seen by science or that are inconsistent or both. For examples, look at religions, especially other than one's own.

A well-known example of that is the biblical description of how the Earth came to be populated by animals. It is described as a divine unique event occurring within a single day. A literal belief in that description, a belief called *creationism*, stands in stark contradiction to the scientific description of the phenomenon as a natural evolutionary process taking millions of years.

Another well-known example is the Christian concept of the trinity, that God is both one and three. That is a clear self-contradiction, since one is not three and three are not one. I understand that the early Christians, who were zealously monotheistic Jews, had much trouble with the concept. They felt it was leading back to the worship of many gods, from which Judaism had strongly distanced itself. Christian theology has learned to live with the contradiction.

Thus personal taste and preference reign supreme in metaphysics, whereas in science their influence, though present, is not of overwhelming importance. In connection with that, compare the amount, intensity, and acrimony of controversy, sometimes developing into personal animosity, in philosophical circles with the paucity, moderation, and civility of controversy in scientific circles. (That is my clear impression of the situation.)

Furthermore and more extremely, I am convinced that many more people throughout history have been killed, tortured, injured, and persecuted because of what are essentially metaphysical positions, such as religion, nationalism, ethnocentrism, and xenophobia, compared with such actions based on truth supported by nature, such as physical survival and protection of gene pool. And I think that holds true even after discounting the cases where the motivation was only *ostensibly* metaphysical. At least that is the way I read history, for what it is worth. You are invited to make your own evaluation.

TRANSCENDENCE AND NONTRANSCENDENCE

Let us compare two types of world view: transcendent world views and nontranscendent world views. The former are world views involving the existence of a reality beyond, or transcending, nature. Nature, with which science is concerned, is viewed as being embedded in, being part of, that transcendent reality. Religions and world views involving supreme beings and creators are of this type, although transcendent world views are not necessarily religions nor do they have to involve supreme beings or creators. For example, one might hold the teleological belief that nature operates toward some end, for the accomplishment of some purpose (which may or may not be known to the believer), without necessarily also believing in the existence of a supreme power bringing that about. Or, one might believe that the universe had a beginning, that it came into being in some sense, without necessarily also believing in a creator. Note that there is no contradiction inherent to scientists' holding transcendent world views. Indeed there are religious scientists. The essential point is that in their scientific work they confine their considerations solely to nature.

Nontranscendent world views, on the other hand, do without any reality transcending nature and make do with nature as all there is. Although denying any reality to deities, creators, and supreme beings, the nontranscendentist does not have to go so far as to deny the existence of mind, consciousness, thought, emotion, feeling, etc. For the nontranscendentist such concepts are conceived of as aspects of the material universe, specifically as activity going on in that extremely complex organ, the brain. The undeniable fact that we are at present far from understanding just how mind, consciousness, and the like are realized by brain activity is no deterrence from holding a nontranscendent world view. The human brain is, after all and as far as I know, the most complex compact system in the universe. And even after cutting out some of the superlative, the human brain, and even the mammalian brain, is indeed unimaginably complex.

The usual objection to that is the following. Brain cells are not conscious. Electric currents do not feel. Neurotransmitters do not think. Electrochemical reactions do not have emotions. So how can mind, consciousness, feeling, and so on be realized by brain activity, which is nothing but the activity of brain cells, neurotransmitters, electric currents, electrochemical reactions, and other such mindless, unfeeling material components? However—and this is the critical point—complex systems can possess properties that are completely irrelevant and meaningless for their component parts. Specifically, brain activity might very well give rise in a natural manner to mind, consciousness, feeling, etc., in spite of the fact that the components of that activity—brain cells, electric currents, neurotransmitters, and so on—do not possess mind, consciousness, and the like. Thought, feeling, etc., would be realizations of the integrated activity of the brain *as a whole*, although they cannot be attributed to any component of that activity.

A well-known example of a complex system possessing properties that are irrelevant and meaningless for its component parts is the following: Consider a few molecules of gas, such as air, confined in a container. The state of those molecules is fully described by their locations and velocities (speeds and directions of motion) at any time. As time goes on the molecules fly about, rebounding from the walls of the container and occasionally, albeit rarely, colliding with each other. Now imagine increasing the number of gas molecules in the container immensely, to many thousands of millions and more. Whatever the number of molecules, their state is still describable, at least in principle, by their locations and velocities at any time, although it might not be practical to actually describe the situation in that way due to the tremendous amount of data involved. And the molecules still fly about, rebounding from the walls of the container, and as their number is increased, colliding with each other more and more frequently.

However, as the number of gas molecules in the container is increased, the situation can better and better be described in macroscopic terms, in terms of volume of gas, its pressure on the container walls, and its temperature. The macroscopic concepts of volume, pressure, and temperature are relevant and meaningful for an ensemble of very many molecules, while they are irrelevant and meaningless for each of the individual molecules the ensemble comprises.

Another well-known example of a complex system possessing properties that are irrelevant and meaningless for its components is an insect colony, such as a colony of ants or bees. An insect colony, very aptly described as a higher-level, or super, organism, possesses the properties of will for survival, adaptation to weather, foresight (preparation for difficult seasons), self-perpetuation (by reproduction), and so on. The behavior of each member of the colony, however, is characterized by automatic reaction to a limited range of tactile, chemical, auditory, and visual stimuli, and an individual insect separated from its colony does not last very long. None of the macroscopic properties of the colony is possessed by any of its members: A soldier will easily get itself killed in defensive or aggressive action of the colony as a whole. A worker might wear itself out bringing food for the colony. An individual might drown while protecting the colony from rain. Only the queen and a drone reproduce at all, and they reproduce the whole colony. The food collected for the winter is intended primarily for the queen and her offspring, so in case of winter shortage a worker might not get to eat any of the food it itself collected. And so on.

Life itself, generally considered by transcendentists to belong to the extranatural domain, is viewed by nontranscendentists as a property that might develop naturally in sufficiently complex systems, in analogy with the above examples, although that possibility is still far from being understood. For each of the individual molecules comprising a living organism, the property of life is irrelevant and meaningless: molecules do not live. Indeed, throughout its lifetime an organism is constantly exchanging molecules with its environment, so any single molecule might be part of the organism for a limited time only.

As we saw above, the nontranscendentist view of mind, consciousness, etc., is similar. They are considered to be properties that might be developed naturally by sufficiently complex systems, not even necessarily living systems. Those properties are irrelevant and meaningless for each of the individual neurons comprising the brain: cells are not considered to be capable of thought or feeling. According to this physicalist point of view, sufficiently complex computers might eventually attain the properties of mind or consciousness. We cannot now know what such computers will be made of, but for whatever will fulfill the function of today's transistor those properties will be irrelevant and meaningless: individual logic elements do not think, feel, and so on.

Although it might eventually be possible to determine whether life, mind, consciousness, etc., are aspects of nature or are extranatural, still, on the whole, there is no objective way of determining the truth or falsehood of world views, as I mentioned above. Thus all world views are *a priori* equally valid, if indeed the concept of validity is relevant at all to world views. As for myself, I would reject any world view that was at odds with science or was inconsistent. But that is a result of my own world view. If someone chooses to ignore science, that might have low survival value for the believer, but by what higher truth is he or she to be judged? And if someone holds inconsistent beliefs, by what higher standard is consistency a virtue?

That is not to say that all world views are equally simple, satisfying, useful, and so on, but those are matters of individual taste and preference. In our present investigation I am not a wholly unbiased participant and am glad to state that my own world view is nontranscendent. (That should not, however, pit me against transcendentists, and there should be no reason for our relations not to be mutually respectful and cordial. I must admit, though, that although I am tolerant of inconsistent beliefs, and probably harbor some of my own, I find it hard to stomach denial of the validity of science.) It seems to me that transcendent world views are intrinsically more complicated than nontranscendent ones in that the former carry the extra baggage of extranaturality. I do not see how transcendentists can deny that. But then they can make the counterclaim that transcendent world views are inherently richer and more meaningful than nontranscendent ones, and that nontranscendentists are denying themselves the richer and more meaningful experience afforded by transcendent world views. And it is hard to argue with that.

However, I do not give up and I respond with the counter-counterclaim that I am far from exhausting the wealth of experience afforded by nontranscendentism and I somehow manage to find enough meaning in it to keep me very happy. Indeed, by the principle of economy (or parsimony), which is part of my own world view, I much prefer making the most of the least, taking minimal conceptual raw material as far as it will go before claiming need for more. And there is still so much to be found in nature that I feel I am wandering in a veritable wonderland. What else could anyone want? Who could ask for anything more? Well, transcendentists do, obviously.

SUMMARY

Metaphysics, in the sense of the philosophical framework in which science operates, is concerned with what lies around, below, above, before, and beyond science. Within the domain of science scientists should stick to scientific considerations as far as possible. Metaphysics should consider only subjects lying outside that domain. World views are conceptual frameworks and lie within the domain of metaphysics. In contrast to science, metaphysics, including world views, lacks criteria for truth, so personal taste reigns supreme. Transcendent world views involve the existence of a reality beyond nature. Nontranscendent world views make do with nature as all there is. Life, mind, consciousness, feeling, etc., can be fitted into nontranscendentism as properties that might develop naturally in sufficiently complex systems. Transcendentism and nontranscendentism are compared.

BIBLIOGRAPHY

Concerning science and religion, see: P. C. W. Davies, *God and the New Physics* (New York: Simon and Schuster, 1983), which discusses also life, mind, consciousness, etc., in chapters 5–9; and J. C. Polkinghorne, *One World: The Interaction of Science and Theology* (Princeton, N.J.: Princeton University Press, 1986). Davies shows that science, specifically physics, can now seriously tackle what were formerly metaphysical questions. He expresses the provocative opinion that science offers a surer path to God than does religion. Polkinghorne, a physicist turned ordained minister, discusses the differences and similarities of science and religion. I cannot resist quoting Polkinghorne's beautifully understated description of Davies on page 77: "Paul Davies, who cannot readily be suspected of being unduly influenced by Christian theology..."

H. Fritzsch, *The Creation of Matter: The Universe from Beginning to End* (New York: Basic Books, 1984), has a discussion of science and religion in chapter 17. See also the "epilogue" of J. S. Trefil, *The Moment of Creation: Big Bang Physics from Before the First Millisecond to the Present Universe* (New York: Charles Scribner's Sons, 1983).

The following two books deal with the development of life from inanimate matter through random processes obeying the laws of nature. Their main import is that the combination of chance and the laws of nature leads to inevitability; no extranatural factors are needed. Related metaphysical issues are discussed.

J. Monod, *Chance and Necessity: An Essay on the Natural Philosophy of Modern Biology* (New York: Knopf, 1971).

M. Eigen and R. Winkler, *Laws of the Game: How the Principles of Nature Govern Chance* (New York: Knopf, 1981).

For a discussion of, among other things, self-organizing systems, including living systems, see: I. Prigogine and I. Stengers, *Order out of Chaos: Man's New Dialogue with Nature* (New York: Bantam Books, 1984), which includes a lot of metaphysical considerations; and chapter 6 of J. D. Barrow, *The World Within the World* (Oxford: Oxford University Press, 1988).

4 Unique Universe

The saga of the universe,
Oft told, but variously. And yet,
Each diff'rent tale does tell its truth
Of Nature's wondrous world.

CAPRICIOUS COSMOS

In chapters 1 and 2 we saw that in doing science we operate as follows: We investigate whatever reproducible aspects of nature interest us and look for order. Upon finding order we formulate laws, which allow prediction. Then we try to develop theories to explain the laws. Thus the raw material of science is the reproducible phenomena of nature; those are the grist for science's mill. If any aspect of nature, however interesting it might be, exhibits no reproducibility, then it will purely and simply lie outside the domain of concern of science, and the concepts of order, law, predictability, and theory will be completely irrelevant to it.

Some examples of more or less known kinds of irreproducible phenomena, if they really exist, are parapsychology, "anomalous events," "transient phenomena," and miracles. Parapsychological phenomena include effects such as extrasensory perception (ESP), telepathy, telekinesis, and clairvoyance. Many controlled experiments have been performed to investigate those effects. To the best of my knowledge the results of the experiments so far indicate that such effects are not absolutely excluded, but, if they exist, they are very weak and have not yet exhibited any reproducibility. In addition, the effects tend to weaken and disappear as the experimental controls are made increasingly rigorous.

Miracles, if indeed one believes in them, are irreproducible by their very nature, so nothing more need be said about them. Those phenomena called "anomalous events" and "transient phenomena" are peculiar, unexpected, unexplained, and definitely irreproducible occurrences. They are like miracles, but are unexpected and seem to have no purpose behind them. For example, one goes to bed at night as usual and wakes up the next morning in a strange bed in a strange city. Or one's TV, all of a sudden and for a few seconds, picks up a broadcast from another country halfway around the world. Ball lightning phenomena are also considered to belong to this category. Who has not heard or read of something along these lines—especially with the help of certain "newspapers" that catch your eye at supermarket checkouts? But again, such phenomena are irreproducible by their very nature.

So even if such interesting but irreproducible phenomena as parapsychology, "anomalous events," and miracles do exist, they will be immune to our *scientific* understanding (although they might be amenable to nonscientific modes of comprehension and understanding). Order, law, predictability, and theory, in the sense of science that we discussed in chapters 1 through 3, will be irrelevant to them, and *in the sense of science* they will justifiably be described as orderless, lawless, unpredictable, and unexplainable.

Note well that the fact that science by its very character cannot comprehend irreproducible phenomena does not preclude their occurrence in nature. One might or might not believe that such phenomena occur in the material universe with which we can, or can conceivably interact, but their incompatibility with the framework of science should not serve as the rationale for anyone's disbelief. Science, as our *attempt* to understand the reproducible and predictable aspects of nature, is not claimed to be the final arbiter of the occurrence of anything.

Stated in other words: Who is to say that all of nature must be reproducible? Not science! Science aims strictly at only those aspects of nature that are reproducible and simply ignores nature's irreproducible aspects. Thus science has nothing whatsoever to say about whether nature possesses any irreproducible aspects or not. For example, telepathy might very well be a real effect. The inability of science to comprehend telepathy due to its irreproducibility is no reason not to be open-minded about its possible existence.

Where are reproducible phenomena found in nature? Or, in other words, where in nature can we search for order, law, predictability, and explanation? We know from experience that sufficiently small systems exhibit reproducible behavior, where sufficiently small is more or less human size and smaller. (Actually, to be more precise, such systems are found to be reproducible when they are isolated from their surroundings in order to prevent the effects of uncontrollable external influences. That is discussed in more detail in the section *Quasi-isolated System and Surroundings* of chapter 5.) Those are the systems we can manipulate and experiment with. As we discussed in the section *Reproducibility* of chapter 1, reproducibility is discovered by repeating experiments. And to be able to experiment and repeat we have to be able to manipulate. As a very ordinary example, we find reproducibility in our own bodies. We can reproducibly tie our shoes, button our buttons, and grasp the objects we intend to. For more precise investigation we perform laboratory experiments, in which we find and study reproducible phenomena.

Systems much larger than human size cannot be manipulated, but such systems might be considered reproducible by *fiat*. For example, geological phenomena, such as earthquakes, mountains, glaciers, and volcanoes, are not reproducible at will in the laboratory. They are, however, treated as being reproducible in the sense that nature supplies us with sufficient variety with sufficient frequency that we can consider nature to be manipulating for us and presenting us with performed experiments at size scales beyond our manipulative capabilities. Such reasoning is extended to even larger systems: planets, satellite systems (planets around a star or moons around a planet), stars, galaxies (groupings of millions of stars), galaxy clusters (groupings of galaxies), and superclusters (clusters of clusters). But the grounds for considering them reproducible weaken toward the end of the list, since the larger the system, the less the variety and the lower the frequency that nature presents us for observation.

As we progress from small, reproducible systems to systems of increasing size, actual reproducibility becomes invalid, to be replaced by declared reproducibility, which in turn loses its justification. When we push matters to their extreme and consider the whole universe, we have clearly and irretrievably lost the last vestige of reproducibility: the universe as a whole is a *unique phenomenon* and as such is inherently irreproducible.

Let us look into that point in more detail. Certainly we cannot set up universes at will. But perhaps the universe can be *declared* reproducible, since theoretical physicists have put forth proposals involving the existence (in some sense) of an unimaginable number of other universes. Well, it cannot thus be made reproducible. As we will see in the following samples, in all those proposed universe ensembles there is no interaction, not even a conceivable possibility of interaction, among universes. So we in our universe cannot, not even conceivably, interact with any of the others. In all those cases, therefore, the proposed other universes are no part of nature and do not interest us in our present investigation. (Recall that nature is the material universe with which *we can, or can conceivably, interact*.) So as far as science is concerned our universe is still a unique phenomenon. Those ensembles are metaphysical ensembles, not natural ones. Here are three samples.

First sample: There is a most successful theory, called quantum theory, that is concerned with the fundamental behavior of all systems in principle, but is usually and most usefully applied to molecular, atomic, and subatomic systems. That theory is formulated in terms of possible happenings and their probabilities of actually occurring (rather than, as for Newton's theory of Kepler's laws, in strictly deterministic terms of what *will* occur). The theory itself is very formal and mathematical, and although it is universally accepted among scientists since it always gives excellent results, its conceptual interpretation is still the subject of much controversy.

One interpretation of quantum theory, called the many worlds interpretation, proposes that at every instant the universe "branches" into realizations of all the quantum possibilities of that instant, which continue to "coexist side-by-side," each branch universe branching further at the next instant, and so on to unimaginable multitude. Thus, by this proposal, there is no fundamental discrimination between what actually happens and what might have happened. Every possibility is realized in its appropriate branch universe, while we happen to find ourselves in only one of the branch universes and perceive reality accordingly. Those "side-by-side, coexisting" universes are supposed by this proposal not to interact with each other. Thus we in our branch universe cannot, not even conceivably, interact with any of the other branch universes. This fact places all the other branch universes outside nature. Thus, the many worlds picture is not a description of nature, but is rather a metaphysical idea. The proposed "side-by-side coexistence" is in a purely metaphysical sense. Science can offer no answer to the question "Where did I win?" asked by the loser of a coin toss.

Second sample: The proposed "cosmic oscillation" scheme of the evolution of the universe has the universe come into being in a "big bang" (a cosmic explosion), expand to maximal extension, contract down again, and end its life in a "big crunch" (a cosmic implosion), followed by the big bang birth of another universe, its expansion, contraction, and final demise in a big crunch, and so on and on, with no beginning and no end. No influence or information is supposed, according to this scheme, to carry over

49

from one universe to the next. This noninteraction among universes makes that ensemble, too, a metaphysical one, since neither can we in our universe interact with any of the others. As far as science is concerned, that "sequence" of universes is no temporal sequence, because ordinary time has meaning only in relation to our universe. It can be thought of as a sequence only in some metatime, a purely metaphysical construct. Questions such as "When is the universe in which I have blue eyes rather than brown?" have no answers within the framework of science.

Third sample: The big bang scheme for the biography of the universe has the universe come into being some 15 thousand million years ago in the form of a cosmic explosion, as was mentioned in the previous paragraph, a primeval fireball of extreme density, pressure, and temperature that has been expanding and cooling ever since. That scheme has a version in which, during an era of unimaginably rapid expansion (called the inflationary era) starting soon after the big bang, the expanding universe broke up into island universes, one of which is ours. Those universes carried on the cosmic inflation by flying away from each other at such stupendous speeds that no interaction among them was possible, since no influence emanating from any island universe could propagate fast enough to overtake any other. They are now supposed to be so far apart and dispersing so rapidly that still no interaction among them is possible or ever will be.

One might feel that this universe ensemble has more of a natural flavor to it than do the previous two. At least I do. I can imagine something like many universes flying apart from each other, while I cannot at all imagine the other two schemes. But the noninteraction again puts the other universes outside the framework of science and renders the ensemble metaphysical, since we in our island universe cannot interact with any of the others. There is no scientific answer even to the question of how far and in what direction our nearest neighbor island universe is supposed to be.

So, being a unique phenomenon, the universe is inherently irreproducible. Thus the universe as a whole lies outside the framework of science. Order, law, predictability, and explanation are irrelevant and inapplicable to it, and we can justifiably call it the orderless, lawless, unpredictable, unexplainable universe—indeed, *the capricious cosmos*. Or, using livelier language, we can say that as far as science is concerned the universe does as it damn well pleases. The universe is the limit of our possible scientific understanding of the material world, while it itself as a whole can have no explanation in science. As for scientific understanding of the working of the universe as a whole, we will never be able to state more than it is because it is.

COSMOLOGY

Cosmology is the study of the working of the cosmos, the universe as a whole. But since the cosmos lies outside the framework of science, it would seem that cosmology is really a branch of metaphysics rather than of science, as it is usually considered to be. The unpredictable, orderless, irreproducible, unique universe certainly exhibits aspects and phenomena that possess reproducibility, order, and predictability. And we certainly can and do explain aspects and phenomena of the universe by other aspects and phenomena of it. As long as cosmology is dealing with the connections and interrelations among those aspects and phenomena, it is a branch of science. But in its holistic mode, when it attempts to comprehend the universe as a whole, as indeed it is basically intended to do, cosmology can only be a branch of metaphysics.

Let us see how cosmology operates. Cosmology attempts to describe the working of the cosmos at present, in the past, and in the future on the basis of all data available to us here and now. Those data are the laws of nature known to us, the present material composition and state of the universe in our cosmic locality (say, the region of the solar system), and information obtained by telescopic means (giving data only about the past, due to the finite speed of light, of radio waves, and of other carriers of information).

But to construct a scheme of the working of the cosmos, we must make assumptions that are completely unverifiable due to the inherent irreproducibility of the universe as a whole.

For example, we might, and invariably do, assume that the laws of nature known to us now were and will be valid also as far into the past and future as our scheme requires. That assumption is good, conservative science practice, but it is unverifiable and is thus a metaphysical assumption. Or, we might, and also invariably do, additionally assume that the laws of nature known to us here in our cosmic locality are also valid everywhere else in the universe. Again, that assumption, although good, conservative science practice, is unverifiable, so that it too is a metaphysical assumption.

Whatever is assumed in those and in many other regards, the most we logically must demand of a scheme of the working of the cosmos is that it be consistent with all the data and that it be self-consistent, i.e., contain no internal contradictions. Now those requirements might not be sufficient to determine a unique scheme, so it might happen that more than one self-consistent scheme of the working of the cosmos fit all the data. There is no guarantee for uniqueness of a consistent scheme, nor does science offer any criterion of preference among consistent schemes. Thus any criterion of preference, such as simplicity, beauty, or conservatism, is necessarily metaphysical. That is true, just as well, of criteria of preference for theories, although we did not mention it in our discussion of theories in chapter 2 (since metaphysics was looked into only in chapter 3).

Cosmological schemes should not be confused with theories, however. Cosmological schemes are concerned with an inherently irreproducible phenomenon, the unique universe, while theories are concerned with laws expressing order among reproducible phenomena. Cosmological schemes are attempts to *describe* the working of the cosmos, not to scientifically *explain* it, as it is inherently scientifically unexplainable. And they are basically metaphysics, since, due to the inherent irreproducibility of the cosmos, they must contain unverifiable assumptions. So if cosmology is basically metaphysics, what is cosmogony, which deals with considerations concerning the *origin* of the universe, if not almost pure metaphysics *a fortiori*?

That is very much *not* to say that cosmology is a waste of time! Schemes of the cosmos, however metaphysical they are and indeed must be, are of immense value for our attempt to understand the universe. They offer insight into the connections and interrelations among the aspects and phenomena that make up the universe as a whole. They guide us in directing our investigations and give us a framework for organizing our data. They supply terms of reference for formulating laws and devising theories. And they might even have predictive power, thus exposing themselves to falsification.

To see that in action consider, for instance, cosmological schemes of the type currently under intensive investigation, the inflationary big bang schemes, briefly outlined in the previous section. Those schemes of the universe are extremely wide ranging, encompassing aspects of the cosmos from the largest scale to the very smallest, from astronomical considerations to those of the elementary particles and their interactions. The schemes describe how those aspects mesh and affect each other, combining to make up the whole. The schemes include effects and phenomena that have not yet been observed, such as the future evolution of the universe, extra-large-scale astronomical structure (involving galaxy clusters and superclusters), new kinds of elementary particles, and new properties of the elementary particle interactions. Thus the schemes guide research in those areas, provide a framework for interpreting and correlating results, and even offer predictions.

Cosmological schemes describe the working of the cosmos and thus, as in the example of the preceding paragraph, retrodict (i.e., tell about the past) the previous evolution and even the origin of the universe, have something to say about its present state, and predict its future evolution. As such they might appear to be laws of behavior for universes. Yet cosmological schemes are not and cannot be laws of behavior for universes, or any other kinds of law for that matter. They were *not* obtained by investigating the behavior of reproducible systems. We cannot run experiments in universe evolution, and nature does not supply us with more than the single universe we have. (At least that is how things seem at present, but we do not see any prospect that the situation will be any different in the future.) Thus there is no reason at all to think that cosmological schemes describe what *really* happened or will happen.

As for the apparent retrodictive power of cosmological schemes, recall that such schemes are based on the available data. Those data in themselves do not tell us the past behavior of the cosmos. Not even do the data obtained by telescopy, although they refer to the past, tell us the universe's past behavior. The data must be interpreted to be meaningful. But it is just such cosmological schemes that serve as tools for interpreting data. And we have no independent way of finding out what really happened. There is no Rip van Winkle who, after sleeping a few billion years, can wake up and tell us what was going on in the universe before he fell asleep. We can only interpret the data we find here and now, and we need cosmological schemes for that.

As an example, consider the astronomical observation known as the red shift effect, which is that the stars in distant galaxies appear redder than those in closer ones. What does that effect tell us about the past, when the light left the stars to travel unimaginably vast distances at the speed of 300,000 kilometers per second until it reached our telescopes? Well, even the idea that the light we receive today is the same light that left the stars is an assumption based on some cosmological scheme telling us what happens to light on its way from stars to us. It could be, for instance, that the original starlight was absorbed by dilute interstellar matter lying in its path, which then emitted new light, redder than the light it absorbed. In that case the red shift would be telling us something about the properties of the interstellar matter at the time the light encountered it.

If we assume, on the other hand, that the light reaching our telescopes is actually the same light that left the stars, then the red shift might be telling us that stars were really redder in the past, since the light reaching us now left the farther stars before it left the nearer ones. Or the red shift might be telling us that stars were always the same color, but that the galaxies are and were moving away from us, with the more distant galaxies receding faster than the closer ones. The reddening of the light would then be the result of the well-known Doppler effect (after Christian Johann Doppler, 1803–53), by which light appears redder or bluer, or a tone sounds lower or higher in pitch, when its source recedes or approaches, respectively. (Recall the change in pitch of an ambulance or police siren as the vehicle speeds by.)

That interpretation of the red shift data is, in fact, the most commonly accepted one. It is based on the cosmological assumption that all the galaxies in the universe are receding from each other, which is consistent with the big bang cosmological scheme, by which the universe has been expanding since it came into being. However, the main point for us is really not this cosmological scheme or that, but rather the fact that, in order to tell us something significant about the past, the data must be interpreted by means of *some* cosmological scheme.

So the apparent retrodictive power of cosmological schemes turns out to be illusory. Cosmological schemes do indeed paint dynamic pictures of the universe's past, perhaps even its origin, on the basis of everything known here and now, but we have no independent source for confirmation. Thus, to put the matter bluntly, as regards telling us the origin of the universe and its past evolution, cosmological schemes are actually no better than fairy tales.

Now cosmological schemes also predict the future evolution of the universe. We have seen that cosmological schemes are not laws, hence there is absolutely no compelling reason to believe that the future they ascribe to the universe is what will really occur. Yet it would seem that simply by waiting patiently it should be possible to compare their predictions with reality. Well, patience will be of no help. The time scales involved in the predicted changes are just so enormous that the predictive power of cosmological schemes, although valid in principle, is utterly nonexistent in practice. To emphasize the point, the Sun can be expected to explode and broil the solar system long before any predicted evolution should become evident. So, as regards telling us the future evolution of the universe, cosmological schemes hardly fare better than when they pretend to tell us its past.

Earlier in this section I stated that cosmological schemes, among their other useful properties, might have predictive power, thus exposing themselves to falsification. In the meantime we have seen that their predictive power involving the past or the future is merely an illusion. The nonillusory remainder is their predictive power regarding the *present* state of the universe. Here lies their big opportunity to shine. Recall that such schemes are based on the available data. They then might predict something about the universe right now, something that can be tested by performing experiments now, obtaining new data, and comparing those data with what the schemes say they should be.

For example, assuming it is not known how galaxy clusters are distributed throughout space (actually that is currently being investigated), a cosmological scheme might predict that the spatial distribution of galaxy clusters is more or less uniform. Then astronomers can perform their observations and analyses and come up with data about galaxy cluster distribution, telling whether they are distributed more or less uniformly or bunched into superclusters. (The latter seems to be the case.) Thus the scheme can be tested. Or, a cosmological scheme might predict the existence of a previously unknown kind of elementary particle or property of elementary particle interaction. Then the particle investigators can get to work and produce data about the existence or nonexistence of the predicted kind of particle or interaction property. And again the scheme can be tested.

So the real predictive power of cosmological schemes, with its concomitant falsifiability, is confined solely to the present. Whatever such schemes have to say about the past and future of the universe should be recognized as illusory from the point of view of science and relegated to the science fiction shelf.

The sweep and range of cosmological schemes can be so exhilarating that some theoretical physicists have expressed their conviction, or at least their feeling, that eventually, with sufficient ingenuity and diligence, it should be possible to develop a Theory of Everything (abbreviated to TOE). As we have learned in the present chapter, a TOE cannot be a theory; it would be a cosmological scheme. It would not be a scientific explanation of the cosmos, but would be a description of it. Yet its descriptions of the origin, past, and future of the universe would be absolutely unfounded from the scientific point of view. Its predictive power would be confined solely to the present state of the universe. Still, were a TOE ever developed, it would certainly be a wonderful accomplishment of immense value, supplying a unifying framework and insight for our attempts to fathom the connections and interrelations among the aspects and phenomena that are components of the universe as a whole. Nevertheless, and to add some perspective, it should be mentioned that there are also those theoretical physicists who believe that a TOE is unattainable either in principle or in practice.

SUMMARY

The universe as a whole is a unique phenomenon as far as science is concerned. So it is inherently irreproducible and thus lies outside the framework of science. Therefore order, law, predictability, and theory are irrelevant to the universe as a whole; it can justifiably be called orderless, lawless, unpredictable, and unexplainable, indeed *the capricious cosmos*. Hence cosmology is basically metaphysics, and cosmological schemes, such as the inflationary big bang schemes, are not theories: they are attempts to *describe*, not scientifically *explain*, the working of the cosmos. Nevertheless, cosmological schemes are of immense value for science, offering insight, guidance, frameworks, and predictions. Such schemes' apparent predictive power concerning the origin and past and future evolution of the universe is illusory, while their real predictive power is confined to the present.

BIBLIOGRAPHY

Here is a list of mostly fairly recent books whose sole or main subject is cosmology. I suggest that a reader try to discern whether the author is assuming that the universe as a whole is subject to law (which would not be science) or not. Also try to note the unverifiable assumptions entering whatever cosmological scheme is being presented. Is the scheme being presented as a scientific theory (which it cannot be)? Is the scheme being presented as The Way Things Were or only as what seems to the author to be the most reasonable interpretation of the available data?

J. D. Barrow and J. Silk, *The Left Hand of Creation: The Origin and Evolution of the Expanding Universe* (London: Heinemann, 1983).

P. C. W. Davies, *The Edge of Infinity* (New York: Simon and Schuster, 1981).

P. C. W. Davies, *The Runaway Universe* (New York: Harper and Row, 1978).

H. Fritzsch, *The Creation of Matter: The Universe from Beginning to End* (New York: Basic Books, 1984).

R. Morris, *The Fate of the Universe* (New York: Playboy Press, 1982).

H. R. Pagels, *Perfect Symmetry: The Search for the Beginning of Time* (New York: Simon and Schuster, 1985, and Toronto: Bantam, 1986).

J. Silk, *The Big Bang*, revised and updated ed. (San Francisco: Freeman, 1989).

J. S. Trefil, *The Moment of Creation: Big Bang Physics from*

Before the First Millisecond to the Present Universe (New York: Charles Scribner's Sons, 1983).

S. Weinberg, *The First Three Minutes: A Modern View of the Origin of the Universe* (New York: Basic Books, 1977).

Now a number of books that are concerned with more than just cosmology, but that contain material on cosmology. My above suggestions hold for them too. See: chapter 16 of R. K. Adair, *The Great Design: Particles, Fields, and Creation* (Oxford: Oxford University Press, 1987); chapter 4 of J. D. Barrow, *The World Within the World* (Oxford: Oxford University Press, 1988); chapter 11 of F. Close, *The Cosmic Onion: Quarks and the Nature of the Universe* (New York: American Institute of Physics, 1983); chapters 2–4, 15, 16 of P. C. W. Davies, *God and the New Physics* (New York: Simon and Schuster, 1983); chapters 5 and 6 of P. C. W. Davies, *Space and Time in the Modern Universe* (Cambridge: Cambridge University Press, 1977); chapters 1–3 of S. W. Hawking, *A Brief History of Time: From the Big Bang to Black Holes* (London: Bantam, 1988); chapters 10–12 of R. Morris, *The End of the World* (Garden City, N.Y.: Anchor Press, 1980); chapter 18 of D. Park, *The How and the Why* (Princeton, N.J.: Princeton University Press, 1988); and chapters 14–15 of J. S. Trefil, *Reading the Mind of God: In Search of the Principle of Universality* (New York: Charles Scribner's Sons, 1989).

For quantum theory (science) and its many worlds interpretation (metaphysics), see: P. C. W. Davies, *Other Worlds: A Portrait of Nature in Rebellion: Space, Superspace and the Quantum Universe* (New York: Simon and Schuster, 1980); P. C. W. Davies and J. R. Brown, eds., *The Ghost in the Atom: A Discussion of the Mysteries of Quantum Physics* (Cambridge: Cambridge University Press, 1986), which includes interviews with physicists actively involved in the foundations of quantum theory; T. Hey and P. Walters, *The Quantum Universe* (Cambridge: Cambridge University Press, 1987), which intentionally does not discuss the philosophy and meaning of quantum theory; J. C. Polkinghorne, *The*

Quantum World (Princeton, N.J.: Princeton University Press, 1984); A. Rae, *Quantum Physics: Illusion or Reality* (Cambridge: Cambridge University Press, 1986); as well as chapter 10 of Adair, above; chapter 3 of Barrow, above; chapter 8 of Davies, *God and the New Physics*, above; chapter 6 of R. P. Feynman, *The Character of Physical Law* (Cambridge, Mass.: MIT Press, 1965); chapters 3 and 4 of Fritzsch, above; chapter 4 of Hawking, above; part 1 of H. R. Pagels, *The Cosmic Code: Quantum Physics as the Language of Nature* (New York: Simon and Schuster, 1982); chapters 15 and 16 of Park, above; and part C of F. Rohrlich, *From Paradox to Reality: Our New Concepts of the Physical World* (Cambridge: Cambridge University Press, 1987).

For other universes (metaphysics), see: chapters 2 and 5 of Barrow and Silk, above; chapter 11 of Davies, *The Runaway Universe*, above; part 3 of J. Gribbin and M. Rees, *The Stuff of the Universe: Dark Matter, Mankind and the Coincidences of Cosmology* (London: Heinemann, 1989); chapter 12 of Morris, *The End of the World*, above; and chapter 8 of Morris, *The Fate of the Universe*, above.

For putative TOEs, Theories of Everything (which would not be scientific theories), see: P. W. Atkins, *The Creation* (San Francisco: Freeman, 1981); P. C. W. Davies and J. R. Brown, eds., *Superstrings: A Theory of Everything?* (Cambridge: Cambridge University Press, 1988), which includes interviews with physicists actively involved in superstring theory, a proposed TOE; chapter 18 of Adair, above; chapters 7 and 11 of Gribbin and Rees, above; and chapter 11 of Hawking, above.

5 Laws of Nature

We are part of all we survey:
Shook and shakers, made and makers,
Seen and seers, kings and subjects.
Ne'er to stand aloof from it,
But neither fully one with it.

In the section *Capricious Cosmos* of the preceding chapter we saw that the universe *as a whole*, being a unique phenomenon, is irreproducible, orderless, lawless, unpredictable, and unexplainable by science. *Within* the universe, however, we find aspects and phenomena that *are* reproducible, orderly, lawful, predictable, and explainable by science. Since aspects of the universe and phenomena within it are parts of the whole universe, the behavior of those parts is part of the behavior of the whole. Their behavior can even be thought of as being engendered by the behavior of the whole. Thus their behavior might be expected to be just as irreproducible, orderless, lawless, unpredictable, and unexplainable as that of the whole. That expectation derives from our common experience that parts of a totally unruly situation are generally every bit as unruly as the whole, if not more so.

Hence arises the question: How is it that within the capricious cosmos there are laws of nature at all? The resolution of that apparent contradiction is the principal subject of the present chapter and the next.

REALISM AND IDEALISM

On the way to our destination, in order to lay a foundation and to broaden understanding, we will pass through and linger a bit at a few stations. Our first stop is a consideration of the question: Where do the laws of nature reside? Nature, at least in certain of its aspects (such as planetary motion), "behaves lawfully," "obeys laws." We discover those laws. Do the laws then reside in our recognition of them? Would the laws not exist if we did not discover them? Still, we would expect nature to behave in the same way whether we discovered its laws or not. Yet could that behavior be "lawful" even if we did not recognize it as such? If it could be, then the laws of nature would be residing in nature external to ourselves. But if there is no way that nature could be considered to obey unrecognized laws, then in some sense the laws would have to reside in ourselves.

So where do the laws of nature reside? Are they wholly "out there," in the world external to the observer? Or are they completely "in here," in the observer's mind? Or some of both? The question is pure metaphysics, of course. The former approach is the metaphysical position called *realism*, which is held by virtually all scientists, although almost always unconsciously and without considering other possibilities. That world view says that the laws of nature are "really" there, independent of observers, that they would have reality even if we were not around to discover them, and that they would be discovered just the same by any kind of intelligence.

The process of discovery, according to the realist position, is "merely" the recognition of the validity of the laws. Thus Newton's laws (see the section *Archetypical Example* of chapter 2), for example, existed and were valid even before Newton discovered, or recognized, them. Realism is part of the spirit of most of science (with an important exception, to be discussed in chapter 8). Although realism does not demand conservatism, it does encourage it. In regard to laws of nature conservatism takes the form: As long as there is no compelling reason to the contrary, assume that the laws of nature we find here and now are, were, and will be valid everywhere and forever.

Regardless of one's political views, conservatism is generally considered to be good science practice. (That is a metaphysical position.) In broad terms the conservative platform in science is: Hold on to what you have, stick to the tried and well confirmed, for as long as is reasonably possible; make changes only when the need for change becomes overwhelming; and then make only the minimal changes needed to achieve the desired end. Thus conservatism demands, among other things, that, given the choice, one prefer theories that assume the laws of nature we find here and now are, were, and will be valid everywhere and forever over theories that assume otherwise. If one held that the laws of nature were not wholly "out there," were not well anchored in the concrete of external reality, one would more likely be tempted, as long as one is theorizing anyway, to assume variable (in time and/or in space) laws of nature than if one held a realist position. Or so it seems to me.

The metaphysical position that the laws of nature are wholly "in here," in the mind of the observer, is called *idealism*. That world view is rare among scientists. Idealism says that the laws of nature we "discover" are not inherent to the "raw material" of the external world, but are mental constructs, artifacts of the way our minds interpret and organize our sensory impressions, of the way we perceive the world. Thus, our laws of nature are determined by the properties of our sense organs and by our mental makeup. Intelligent beings much different from us might be able to detect aspects of the universe we cannot even imagine or might have no sense for light, sound, or odor, for example. Then their different mental makeup might interpret and organize their different sensory impressions so as to "discover" laws of nature so wildly different from ours, yet valid for the world as they perceive it, that their laws would be irrelevant to the world as we perceive it. We might very well wonder if they and we would be able to communicate at all or even recognize intelligence in each other.

Between the poles of realism and idealism lies a range of possibilities for other metaphysical positions concerning the seat of the laws of nature. One such a hybrid world view, which I find especially attractive, is that there indeed is order "out there" in the world external to ourselves and that intelligent beings perceive it via their limited senses and formulate laws from those perceptions in accord with their mental makeups. Thus order would be objective, while laws of nature would be species-subjective, perhaps even individual-subjective.

On the one hand, by that position, intelligent beings evolve and develop within and by the objective order, so their senses and mental makeups must have survival value and cannot bring about behavior that is too "unnatural," such as acting according to "laws" that have no objective support or ignoring orderly phenomena that affect the beings strongly.

As an example of acting according to nonlaws, imagine a being on Earth who somehow forms the "law" that whenever a light source is present in the sky (Sun, Moon, or stars) a tornado will strike imminently and it is essential to bury one's head in the sand to avert danger. Such a being would have a very low chance of surviving, mainly since there would remain little time for it to look for food, and secondarily since it would not be very well protected should a tornado actually happen to strike.

For an example of ignoring important objective order, consider an imaginary being who searches for food at random times during the day and night, ignoring the fact that food is available only during the day. Such a being would both waste energy searching for nonexistent food at night and not fully exploit the food potential of the day, thus suffering a disadvantage compared to a being who acted in better accord with the objective order.

Additional, human examples of those two antisurvival behaviors are astrology (acting according to nonlaws) and rejection of the insights of medical science (ignoring objective order). In the former case, a fanatically astrology-believing general would choose to give up a tactically favorable, but horoscopically unfavorable, date for attack and instead prefer to move his or her forces on a horoscopically favorable, but tactically unfavorable, date. The result could be catastrophic (for that side of the conflict). Fortunately for them, most astrology-inclined people act according to their horoscope only in matters where it makes little difference either way.

Rejecting the insights of medical science is often a serious matter. By ignoring the well-known dangers of smoking, smokers reduce both the quality of their lives and their life expectancies. Eaters of poor diets do the same, when they ignore what is known about the effect of diet on health. Whatever the advantages of alternative methods of healing might be, categorical rejection of conventional medicine is often a source of suffering and death.

On the other hand, although, as we just saw, the objective order "out there" must impose certain features on the laws of nature formed by any viable species or individual, the hybrid position we are considering recognizes that differing senses and mental make-ups can certainly develop. That could happen when beings evolve in diverse environments, with the formation of consequently differing world views and laws of nature (within the restrictions imposed by the objective order). Compare, for example, what we can reasonably infer to be the laws of nature of dolphins with those of birds, or each with ours. And, letting our imagination wander, we would hardly expect an intelligent galaxy, should a galaxy have or ever develop intelligence, to view matters in quite the same manner as we do.

REDUCTIONISM AND HOLISM

The next stop on our way to a clarification of how it is that there are laws of nature at all is yet another polarity, additional to the realism-idealism polarity.

We live in nature and view it and are intrigued. Our material needs and our curiosity drive us to try to understand what is happening around us, in order both to improve our lives by better satisfying our material needs and to satisfy our curiosity. What we observe in nature is a complex of phenomena—some apparently interrelated, others seemingly independent—including ourselves, where we are related to all of nature, as is implied by our definition of nature as the material universe with which *we can, or can conceivably, interact*. That possibility of interaction is what relates us to all of nature and, due to the mutuality of interaction and of the consequent relation, relates all of nature to us.

It then follows that all aspects and phenomena of nature are actually interrelated, whether they appear to be so or not. Whether they are interrelated independently of us or not, they are certainly interrelated through our mediation. For example, let us say that aspect A of nature is not directly related to phenomenon P. But since both A and P are components of nature, we can interact with both. Thus they are interrelated by means of us. A can interact with P by interacting with us and thereby causing us to interact with P, and *vice versa*. In that manner all of nature, including *Homo sapiens*, is interrelated and integrated.

Science is our attempt to understand (the reproducible and predictable aspects of) nature. But how are we to grasp that wholeness, that integrality? When we approach nature in its completeness, it appears so awesomely complicated, due to the interrelation of all its aspects and phenomena, that it might seem utterly beyond hope to understand anything about it at all. True, some obvious simplicity stands out, such as day-night periodicity, the annual cycle of the seasons, and the fact that fire consumes. And subtler simplicity can be discerned, such as the term of pregnancy, the correlation between clouds and rain, and the relation between the tide and the phase of the Moon. Yet, on the whole, complexity seems to be the norm, and even simplicity, when considered in more detail, reveals wealths of complexity. But, due to nature's unity, to its integrality that we just demonstrated, any attempt to analyze nature into simpler component parts cannot but leave something out of the picture.

That brings us to the metaphysical position called *holism*. According to the holist world view, nature can be understood only in its wholeness or not at all. And that includes human beings as part of nature. As long as nature is not yet understood, there is no reason *a priori* to consider any aspect or phenomenon of it as being intrinsically more or less important than any other. Thus it is not meaningful to pick out some part of nature as being more "worthy" of investigation than other parts. Neither is it meaningful, according to the holist position, to investigate an aspect or phenomenon of nature as if it were isolated from the rest of nature. The result of such an effort would not reflect the normal behavior of that aspect or phenomenon, since in reality it is not isolated at all, but is interrelated with all of nature, including ourselves.

The other pole is the metaphysical position of *reductionism*. The reductionist position is that nature is indeed understandable as the sum of its parts, that nature should be studied by analysis, should be "chopped up" into, or reduced to, simpler component parts that can be individually understood. A successful analysis should then be followed by synthesis, whereby the understanding of the parts is used to help attain understanding of larger parts compounded of understood parts. If necessary, that should then be followed by further synthesis, further compounding of the compound parts to obtain even larger parts. Then an understanding of the latter can be attained with the help of the understanding achieved so far. And so on to the understanding of ever-larger parts, until an understanding of all of nature is reached.

To compare the two viewpoints, consider the phenomenon of life. A reductionist would say that a living organism can be understood by understanding the workings of its smallest components and then gaining knowledge of progressively larger components by building up from each level to the next larger one. We should start by investigating atoms and how they combine to form molecules. Then we should investigate how molecules combine to form macromolecules. Then how the latter combine to form cell components, such as membranes. And finally how such components combine to form a whole cell. Then we would understand the living cell.

A holist, on the other hand, would claim that a living cell is more than merely the sum of its components, that life is a property of the cell *as a whole* and cannot be understood in terms of parts. The most that can be understood by the reductionist method, a holist would claim, would be a dead cell.

Between the poles of reductionism and holism lies a range of positions with regard to the separability of nature. While it is debatable if and to what extent one's position along the realism-idealism range affects the way one does science (and indeed there are arguments in both directions), it is clear that one's position within the reductionism-holism polarity should strongly influence the way one does science and even whether one does science at all. Now, each of the poles of holism and reductionism has a valid point to make. Nature is certainly interrelated and integrated, at least in principle, and we should not lose sight of that fact. But if we hold fast to extreme holism, everything will seem so frightfully complicated that it is doubtful whether we will be able to do much science.

On the other hand, separating nature into parts seems to be the only way to search for simplicity within nature's complexity. But a position of extreme reductionism might also not allow much science progress, since nature might not be as amenable to separation into parts as that position claims. So while one's position along the realism-idealism range, such as my own hybrid position stated above, can really be justified only subjectively, by one's finding it attractive or satisfying, it seems best to determine one's reductionism-holism position by the purely pragmatic criterion of whether it allows, or even encourages, science to progress or not. We will now consider three ways nature is commonly sliced up by science and the reductionism-holism position implied by each.

OBSERVER AND OBSERVED

Reduction of nature to its simpler parts can be carried out in many different ways. As the old saying goes, there's more than one way to slice a salami. The most common way of analyzing nature is to separate it into two parts: the observer—us—and the observed—the rest of nature. That separation is so obvious that it is often overlooked. It is so obvious, because in doing science, in our attempt to understand the reproducible and predictable aspects of nature, we *must* observe nature to find out what is going on and what needs to be understood.

Now what is happening is this: We and the rest of nature are in interaction, as was pointed out earlier; observation is interaction. Thus anything we observe inherently involves ourselves too. The full phenomenon is thus at least as complicated as *Homo sapiens*. Every observation must include the reception of information by our senses, its transmission to our brain, its processing there, its becoming part of our awareness, its comprehension by our consciousness, etc. We appear to ourselves to be so frightfully complicated that we should then renounce all hope of understanding anything at all.

So we separate nature into us, on the one hand, and the rest of nature, on the other. The rest of nature, as complicated as it might be, is much less complicated than all of nature, since our own complexity has been taken out of the picture. We then concentrate on attempting to understand the rest of nature. (We also might, and indeed do, try to understand ourselves. But that is another story.) However, referring to the discussion of holism in the preceding section, since nature with us is not the same as nature without us, what right have we to think that any understanding we achieve by our observations is at all relevant to what is going on in nature when we are not observing?

The answer is that in principle we simply have no such right *a priori*. What we are doing is *assuming*, or adopting the working hypothesis, that the effect of our observations on what we observe is sufficiently weak or can be made so, that what we actually observe well reflects what would occur without our observation, and that the understanding we reach under that assumption is relevant to the actual situation. That assumption might be a good one or it might not, its suitability possibly depending on the aspect of nature that is being investigated. It is ultimately assessed by the degree of its success in allowing us to understand nature.

Clearly the observer-observed analysis of nature is very successful in many realms of science. One example is Newton's explanation of Kepler's laws of planetary motion, presented in the section *Archetypical Example* of chapter 2. That excellent understanding of an aspect of nature was achieved under the assumption that observation of the planets does not affect their motion substantially. In general, the separation of nature into observer and observed seems to work very well from astronomical phenomena down through ordinary-size phenomena and on down in size to microscopic phenomena.

However, at the microscopic level, such as in the biological investigation of individual cells, extraordinary effort must be invested to achieve a good separation. The ever-present danger of the observation's distorting the observed phenomena, so that the observed behavior does not well reflect the behavior that would occur without observation, must be constantly circumvented. That is accomplished in the case of cell research, for example, by the use of microelectrodes and micropipettes for investigating the electrical and chemical aspects of cell behavior.

But at even smaller, at submicroscopic sizes, at the molecular, atomic, and nuclear levels and at the level of the so-called elementary particles and their structure, the observer-observed separation of nature does not work. Here it is not merely a matter of lack of ingenuity or insufficient technical proficiency in designing devices that minimize the effect of the observation on the observed phenomena. Here we find that the observer-observed interrelation cannot be disentangled *in principle*, that nature absolutely forbids our separating ourselves from the rest of itself.

Quantum theory is the branch of science that successfully deals with such matters. From it we learn that nature's observer-observed disentanglement veto is actually valid for *all* phenomena of *all* sizes. Nevertheless, the *amount* of residual observer-observed involvement, after all efforts have been made to separate, can be more or less characterized by something like atom size. Thus an atom-size discrepancy in the observation of a planet, a house, or even a cell is negligible, while such a discrepancy in the observation of an atom or an elementary particle is of cardinal significance.

The observer-observed nonseparability is one aspect of the general nonseparability of nature that quantum theory describes. Nonseparability is not easy to grasp and takes a lot of getting used to. It is very counterintuitive, because we have nothing like it at the level of ordinary-size phenomena. We are ordinarily used to the idea that, if we have object A here and object B there, then A is clearly here and B is clearly there and they are well separated. However, at very small sizes we have to be more precise about what we mean. Object A being here means that A is confined to some region of space labeled "here," and similarly for B being "there." A and B being separated means that the two regions of space do not overlap.

But quantum theory teaches us that nature abhors the confinement of submicroscopic objects, such as atoms and elementary particles, to very small regions of space. (That is actually true for objects of all sizes, but the effect becomes negligible for increasing size.) The smaller the confinement region, the sooner the object will escape it and the faster it will move elsewhere! So if we confine A and B, say electrons, to very small regions to ensure good separation, they will both very soon be anywhere else, and we will have failed to keep them apart. If, on the contrary, we confine them to very large regions in the hope they will stay within their assigned regions, the regions will be overlapping, and we will have failed again. And it turns out the same for intermediate-size regions as well. We just can't win! Nonseparability is a general and inescapable fact of nature and cannot be circumvented.

With regard to the reductionism-holism view of observer-observed separation, the most useful position to hold in the range between the poles—useful in the sense of allowing the best grasp of the situation—seems to be determined by the size of the observed phenomenon: the larger the size, the closer to the reductionism pole; while the smaller the size, the closer to the holism pole. The larger the observed phenomenon, the more immune it is to the effects of our observations; while the smaller the observed phenomenon, the more susceptible it is to those effects and the more care has to be taken by observers to minimize those effects. For molecular phenomena and smaller, nearly pure holism seems to be appropriate, since nature's inherent nonseparability is highly significant for them.

QUASI-ISOLATED SYSTEM AND SURROUNDINGS

Whenever we separate nature into observer and the rest of nature, we achieve simplification of what is being observed, because, instead of observing all of nature, we are then observing only what is left of nature after we ourselves are removed from the picture. Yet even the rest of nature is frightfully complicated. That might be overcome by the further slicing of nature, by separating out from the rest of nature just that aspect or phenomenon that especially interests us. For example, if we are interested in hydrogen atoms, we might detach one or more hydrogen atoms from the rest of nature and focus our observations on them, while ignoring what is going on around them. Or, in order to study liver cells

Laws of Nature

we might remove a cell from a liver and examine it under a microscope. This further separating of nature was actually tacitly implied in the previous section, where we considered the investigation of phenomena rather than of all of the rest of nature.

But what right have we to think that by separating out a part of nature and confining our investigation to it, while completely ignoring the rest, we will gain meaningful understanding? We have in principle no right at all *a priori*. Ignoring everything going on outside the object of our investigation will be meaningful if the object of our investigation is not affected by what is going on around it, so that it really does not matter what is going on around it. That will be the case if there is no interaction between it and the rest of nature. Such a system, one that does not interact with its surroundings, is called an *isolated system*.

Now, an isolated system is an idealization. By its very definition we cannot interact with an isolated system, so no such animal can exist in nature, where nature is, we recall, the material universe with which *we can, or can conceivably, interact*. Well then, how about a system that is observable, i.e., is not isolated from us, yet is isolated from the rest of nature? The state of our present understanding of nature, as incomplete as it might be, is still sufficient to answer that question in the negative. For instance, no system can be strictly isolated from the gravitational influence of its surroundings. The gravitational force between two bodies decreases in magnitude with increasing separation, but for no separation does that force absolutely vanish. And no way is known to screen out the gravitational force. So as nearly isolated as a system might be, it will still be subject to gravitational influence from the rest of nature. The best we can do is to attempt to minimize that influence by removing the object system as far away as possible from other bodies.

Another kind of nonisolability has to do with *inertia*. Inertia is the name given to the property according to which a body's behavior is governed by Newton's first universal law of motion (see the section *Archetypical Example* of chapter 2), that in the absence of forces acting on them or when such forces cancel each other, bodies remain at rest or continue to move uniformly in a straight line. Or expressed positively, inertia is the property of bodies that force is required to alter their state of motion from one of rest or of uniform straight-line motion. The state of rest, by the way, is but a special case of uniform straight-line motion—one with constant zero velocity. (It might be mentioned that inertia can be made a measurable physical quantity by means of Newton's second law, according to which it turns out that a body's mass serves as a good measure of its inertia. But that point is not directly relevant to our present discussion.)

Now, motion that seems uniform and straight-line to one observer might very well not appear that way to another. For example, if the motion of an aircraft appears to me uniform and straight-line as I watch it standing on the sidewalk, the same motion could not possibly appear that way to you while you run around me in circles. By Newton's first law I deduce that all the forces acting on the aircraft cancel each other out completely. But if that is true (and indeed it is), then Newton's first law cannot be valid for you too, since you perceive nonuniform and non-straight-line motion. But if you assume that Newton's first law is valid for you, then you must deduce that the forces acting on the aircraft do not completely cancel out, so that Newton's first law could then not be valid also for me.

Hence it is meaningful to ask: For whom is Newton's first law valid? Relative to what does Newton's first law hold? Or, relative to what is force-free motion uniform and straight-line? And to the best of our present understanding the answer seems to be: Relative to all the matter in the universe. Or equivalently, relative to the distant galaxies, where practically all of the matter seems to reside. (And not, strictly speaking, relative to the Earth, since the Earth is undergoing rather complex motion relative to the distant galaxies, as mentioned in the section *Reproducibility* of chapter 1. Nevertheless, for many purposes, and certainly for everyday needs, the

Earth can be used as a sufficiently good reference for Newton's laws.)

That being the case, Ernst Mach (1838–1916) proposed what has come to be known as the *Mach principle*, which is that the origin of inertia lies with all the matter of the universe, i.e., that the inertia of any body is due to all the other matter present in the universe. Another way of expressing that, an expression possessing reductionist character, is that the inertia of any body is caused by some influence, some interaction, between the body and all the other matter in the universe. About the nature of that Machian influence we know virtually nothing. But since we know of no way to diminish, let alone abolish, the inertia of bodies, it is clear that even for the most nearly isolated systems the Machian influence must still be in full force.

On the other hand, the Mach principle might possibly be better grasped as a holistic principle involving all the matter of the universe, which might better not be thought of as separable bodies with some inertia-causing influence among them. In either case that adds another anti-isolatory factor to the gravitational influence. Still, with Newton's help we do know something about inertia. (By the way, Einstein was much influenced by the Mach principle when he developed his general theory of relativity.)

It is important to point out that as long as the Mach principle is not realized in a manner that science can dig its teeth into, it must be taken as a purely metaphysical principle, definitely not a scientific theory. Indeed, experimentation so far has revealed nothing that might indicate the possibility of explaining inertia scientifically. Thus the Mach principle must be considered a guiding principle, offering us a framework for organizing our science concepts, and in no way should it be understood as a theory for the explanation of inertia.

Of the other influences that we are aware of, only the forces of electricity and magnetism possess sufficient range to be potential obstacles to the isolation of systems. The properties of those forces are well known. They both weaken with increasing separation, and, in contrast to the force of gravitation, they can both be effectively screened out. That leaves the nuclear forces (the strong and the weak), which are of such short range that they do not hinder isolation.

There does exist an additional anti-isolatory factor, however, although "influence" might not be a good description of it. It is the intrinsic nonseparability of nature as described by quantum theory, which we discussed in the previous section. Its implication for the isolability of systems is that submicroscopic aspects of systems might be linked to submicroscopic aspects of their surroundings in a holistic manner. The linkage manifests itself through correlations, or interconnections or correspondences, between happenings within the system and happenings outside it. It is not presently very clear whether that linkage can usefully be viewed as some kind of mutual influence or not. In any case it is uncontrollable, and it can be neither screened out nor attenuated by large separation.

So the situation with regard to isolated systems is as follows. To enable themselves to do science, scientists reductionistically try to investigate relatively simple isolated systems. But "isolated" systems are never really totally isolated. They are not totally isolated from us, otherwise we could not observe them (and they would not be part of nature). Neither are they totally isolated from the rest of nature.

The short-range influences—the nuclear forces—are negligible at distances larger than submicroscopic. And the long-range influences—the gravitational, electric, and magnetic forces—can be made as weak as desired by sufficiently removing the system from other matter, while the latter two forces can even be screened out. The putative Machian influence responsible for inertia, if indeed the Mach principle should be viewed that way rather than holistically, cannot be prevented, however, but the properties of inertia are known and can be taken into account. Nevertheless, the quantum correlations that link systems with their surroundings, in manifestation of nature's inherent nonseparability, are uncontrollable and unavoidable. They impose an insurmountable limit to isolability and hence also to the usefulness of reductionism, a limit that scientists have no choice but to live with and to learn to comprehend holistically.

And those are only the factors we are aware of; the existence of additional, unknown, possibly anti-isolatory factors cannot be precluded. Therefore we will henceforth use the term *quasi-isolated system* for an "isolated" system, i.e., for a system that is as nearly isolated as possible.

INITIAL STATE AND LAW OF EVOLUTION

The previous two ways of analyzing nature into parts—separation into observer and the rest of nature and separation into quasi-isolated system and its surroundings—are literal applications of the reductionist position. The present way of analyzing is a metaphoric application, or a generalization of the idea of a part of nature. Rather than a separation that can usually be envisioned spatially—observer here, observed there, or quasi-isolated system here, its surroundings around it—the present analysis is a conceptual separation, the separation of natural processes into initial state and law of evolution.

Things happen. Events occur. Changes take place. *Nature evolves.* That is the relentless march of time. The process of nature's evolution is of special interest to scientists, since predictability, one of the cornerstones of science, has to do with telling what will be in the future, what will evolve in time. Nature's evolution is certainly a complicated process. Yet reproducibility, order, law, and predictability can be found in it, when it is properly sliced. First the observer should separate himself or herself from the rest of nature, as discussed in the section *Observer and Observed* earlier in this chapter. Then he or she should narrow the scope of investigation from all of the rest of nature to quasi-isolated systems and investigate the natural evolution only of such systems, as discussed in the preceding section. Actually, as was mentioned in the section *Capricious Cosmos* of chapter 4, it is only for quasi-isolated systems that reproducibility, order, law, and predictability are found.

Finally, and this is the present point, the natural evolution of quasi-isolated systems should be analyzed in the following manner. The evolution process of a system should be considered as a sequence of *states* in time, where a state is the condition of the system at any time. (That sequence might be continuous or discrete, i.e., smoothly varying in time or jumpy.) For example, the solar system evolves, as the planets revolve around the Sun and the moons revolve around their respective planets. Now imagine that some duration of that evolution is recorded on a reel of photographic film or on a videocassette. Such a recording is actually a sequence of still pictures. Each still picture can be considered to represent a state of the solar system, the positions of the planets and moons at any time. The full recording, the reel or cassette, then represents a segment of the evolution process. (Although the recording happens to be a discrete sequence that jumps from one frame to the next, the actual natural process is continuous, varying smoothly in time.)

Then the state of the system at every time should be considered as an *initial state*, a precursor state, from which the following remainder of the sequence develops, from which the subsequent process evolves. For the solar system, for instance, the positions of the planets and moons at every time, such as when it is twelve o'clock noon in Tel Aviv on 20 October 1990, say, or any other time, should be considered as an initial state from which the subsequent evolution of the solar system follows.

When that is done, when natural evolution processes of quasi-isolated systems are viewed as sequences of states, where every state is considered as an initial state initiating the system's subsequent evolution, then it turns out to be possible to find reproducibility, order, law, and predictability. What turns out is that, with a good choice of what is to be taken as a state for any quasi-isolated system, one can discover a law that, given *any* initial state, successfully predicts the state that evolves from it at *any* subsequent time. Such a law of nature, since it is specifically concerned with evolution, is referred to also as a *law of evolution*.

For an example let us return to the solar system. It turns out that the specification of the positions of all the planets and moons at any single time is insufficient for the prediction of their positions at later times. Thus the specification of states solely in terms of position is not a good one for the purpose of finding lawful behavior. However, the description of states by both the positions and the velocities (i.e., the speeds and directions of motion) of the planets and moons at any single time does allow the prediction of the state evolving from any initial state at any subsequent time. The law of evolution in that case consists of Newton's three universal laws of motion and law of universal gravitation (see the section *Archetypical Example* of chapter 2). It operates with much computer number crunching (huge amounts of numerical calculation) and is indeed capable of predicting the positions and velocities of the planets and moons for any time, given their positions and velocities at any earlier time.

In fact, there are even laws of evolution, and Newton's laws are among them, that also work in reverse. For any initial state they not only predict the state that will evolve from it at any later time, but also retrodict the state that occurred at any earlier time and subsequently evolved into the given initial state, which for retrodiction might then more accurately be referred to as a final state. Thus, given the positions and velocities of the planets and moons at any time, Newton's laws allow the calculation of their positions and velocities for any other time, both earlier or later. Newton's laws, moreover, serve as the law of evolution not only for the solar system, but for any quasi-isolated system of bodies interacting with each other solely through the gravitational force.

So the analysis needed to enable the discovery of reproducibility, order, law, and predictability in the natural evolution of quasi-isolated systems is the conceptual splitting of the evolution process into initial state and law of evolution. The usefulness of such a separation depends on the independence of the two "parts," on whether for a given system the same law of evolution is applicable equally to any initial state and whether initial states can be set up with no regard for what will subsequently evolve from them. Stated in other words, the analysis of the evolution process into initial state and law of evolution will be useful if, on the one hand, nature indeed allows us (at least in principle) complete freedom in setting up the initial state, while, on the other hand, what evolves from an initial state is entirely beyond our control. Only then will that analysis be useful, because only then will the two separated "parts" be independent, and only then will we have thereby achieved simplification and found order and law.

That reductionist analysis of evolution processes into initial states and laws of evolution has proved to be admirably successful for ordinary-size quasi-isolated systems and has served science faithfully for ages. That is how reproducibility is discovered: Set up the "same" initial state in the "same" apparatus and obtain the "same" result. That also gives predictability: A law of evolution predicts the outcome of any relevant initial state. That is the foundation of all laboratory experiments. We rely on that analysis in whatever we do: We insert and turn a key, confident that the door will become unlocked. We move our legs in a certain way and fully expect to find ourselves walking.

The extension of that analysis to the very small seems quite satisfactory, although when quantum theory becomes relevant, the character of an initial state becomes quite different from what we are familiar with in larger systems. Its extension to the large, where we cannot actually set up initial states, is also successful. (See the section *Capricious Cosmos* of chapter 4.) But we run into trouble when we consider the universe as a whole. The principal reason is (as we saw in that section of chapter 4) that the concept of law is irrelevant to the universe as a whole. Although the universe clearly evolves, in its entirety it is in principle not subject to law. (It might be mentioned that also the concept of a beginning, of a literally initial state, is meaningless for the universe as a whole. We do not elaborate on that point, but related topics are discussed in chapter 9.) So, with regard to the possibility of splitting its evolution into initial state and law of evolution, the universe as a whole is shaking its head and commanding us to grasp the situation holistically.

SUMMARY

The metaphysical position of realism is that the order and laws of nature we find are really "out there," independent of observers. Idealism, on the other hand, holds that order and laws of nature are wholly in the mind of the observer. A possible hybrid position is that order is an objective property of nature, while laws are mental constructs. Holism is the world view that nature can be understood only in its wholeness or not at all. The opposite position of reductionism is that nature is understandable as the sum of its parts and should be studied by analysis and synthesis. One way that science reduces nature to its parts is the observer-observed separation. It works well down to the limit set by nature's inherent nonseparability, i.e., for phenomena that are not too small. Another way of reducing nature to parts is the separation into quasi-isolated system and its surroundings, where a quasi-isolated system is a system that is isolated from its surroundings to the best of our ability and understanding, albeit imperfectly due to inertia and the uncontrollable quantum correlations involved in nature's nonseparability. Nature's order is manifested in quasi-isolated systems, and it is for them that laws of nature are found. A third way that science reduces nature is the analysis of the evolution of quasi-isolated systems into initial state and law of evolution.

BIBLIOGRAPHY

For realism-idealism, see: B. d'Espagnat, *In Search of Reality* (New York: Springer-Verlag, 1983), which is not too easy reading; E. Harrison, *Masks of the Universe* (New York: Macmillan, 1985); R. Morris, *Dismantling the Universe: The Nature of Scientific Discovery* (New York: Simon and Schuster, 1983); as well as chapter 1 of R. K. Adair, *The Great Design: Particles, Fields, and Creation* (Oxford: Oxford University Press, 1987); chapter 16 of M. Eigen and R. Winkler, *Laws of the Game: How the Principles of Nature Govern Chance* (New York: Knopf, 1981); chapter 16 of H. Fritzsch, *The Creation of Matter: The Universe from Beginning to End* (New York: Basic Books, 1984); the "epistemological prolegomena" of K. Lorenz, *Behind the Mirror: A Search for a Natural History of Human Knowledge* (New York: Harcourt Brace Jovanovich, 1977); part 3 of R. Morris, *The Nature of Reality* (New York: McGraw-Hill, 1987); chapter 3 of D. Park, *The How and the Why* (Princeton, N.J.: Princeton University Press, 1988); chapter 10 of A. Rae, *Quantum Physics: Illusion or Reality* (Cambridge: Cambridge University Press, 1986); chapter 11 of J. S. Trefil, *Reading the Mind of God: In Search of the Principle of Universality* (New York: Charles Scribner's Sons, 1989); and chapter 12 of A. Zee, *Fearful Symmetry: The Search for Beauty in Modern Physics* (New York: Macmillan, 1986).

For reductionism-holism, see: chapter 5 of P. C. W. Davies, *God and the New Physics* (New York: Simon and Schuster, 1983); chapter 4 of J. Monod, *Chance and Necessity: An Essay on the Natural Philosophy of Modern Biology* (New York: Knopf, 1971); and the "epilog" of J. S. Trefil, *The Moment of Creation: Big Bang Physics from Before the First Millisecond to the Present Universe* (New York: Charles Scribner's Sons, 1983).

For observer-observed, see chapter 10 of Adair, above.

For quantum nonseparability (an anti-isolatory factor), see: chapter 10 of Adair, above; chapter 3 of J. D. Barrow, *The World Within the World* (Oxford: Oxford University Press, 1988); chapter

8 of Davies, *God and the New Physics*, above; chapter 4 of P. C. W. Davies, *Other Worlds: A Portrait of Nature in Rebellion: Space, Superspace and the Quantum Universe* (New York: Simon and Schuster, 1980); chapter 1 of P. C. W. Davies and J. R. Brown, eds., *The Ghost in the Atom: A Discussion of the Mysteries of Quantum Physics* (Cambridge: Cambridge University Press, 1986); chapters 11–13 of part 1 of H. R. Pagels, *The Cosmic Code: Quantum Physics as the Language of Nature* (New York: Simon and Schuster, 1982); chapters 7 and 8 of J. C. Polkinghorne, *The Quantum World* (Princeton, N.J.: Princeton University Press, 1984); chapters 3–5 of Rae, above; and chapter 10 of F. Rohrlich, *From Paradox to Reality: Our New Concepts of the Physical World* (Cambridge: Cambridge University Press, 1987).

For the forces of nature (more anti-isolatory factors), see: P. C. W. Davies, *The Forces of Nature*, 2d ed. (Cambridge: Cambridge University Press, 1986); chapters 12–15 of Adair, above; chapter 4 of F. Close, *The Cosmic Onion: Quarks and the Nature of the Universe* (New York: American Institute of Physics, 1983); chapter 1 of P. C. W. Davies, *The Accidental Universe* (Cambridge: Cambridge University Press, 1982); chapter 5 of S. W. Hawking, *A Brief History of Time: From the Big Bang to Black Holes* (London: Bantam, 1988); chapter 2 of Morris, *The Nature of Reality*, above; chapter 6 of part 2 of Pagels, above; and chapter 5 of Trefil, *The Moment of Creation*, above.

For the Mach principle (yet another anti-isolatory factor), see: chapter 13 of Adair, above; and chapter 3 of J. Silk, *The Big Bang*, revised and updated ed. (San Francisco: Freeman, 1989).

For adaptive biological evolution, see: Lorenz, above; chapter 5 of Eigen and Winkler, above; and chapter 7 of Monod, above.

6 Laws of Nature (Continued)

Things are as they are because that's how they are.
Says nothing, yet says what can be said,
When intellect reaches the end of its scope
In the only world we know.

We continue from the preceding chapter along the way toward resolution of the apparent contradiction in the existence of laws of nature within the capricious cosmos.

EXTENDED MACH PRINCIPLE

Quasi-isolated systems exhibit reproducibility, order, law, and predictability. And the same laws are found to be obeyed by all systems to which the laws are applicable. As an example, Newton's laws are found to hold for all quasi-isolated astronomical systems of bodies (which indeed interact solely through the gravitational force), be they galaxy clusters, single galaxies, binary systems (pairs of stars that are gravitationally closely bound and revolve around each other, a rather common astronomical phenomenon), stars and their planetary systems (like the Sun and the solar system), or planets and their moon systems (like the Earth and its single moon or Jupiter and its many moons). So we are approaching the destination toward which we set out at the beginning of the preceding chapter, which is the resolution of the apparent contradiction implied in the question: How is it that within the capricious cosmos there are laws of nature at all?

But before looking into that question, there still remains one other station at which we should stop. Before considering the confrontation of the order in the behavior of quasi-isolated systems with the inherent orderlessness of the universe as a whole, we should consider the origin of the very behavior of quasi-isolated systems: How do quasi-isolated systems "know" how to behave at all? At every moment how does a quasi-isolated system "know" what to do next? How does it constantly obey the *same* laws? How do the system of the planet Jupiter and its moons and some binary star system located thousands of light-years from us both obey the *same* laws (including Newton's laws)?

Let us recall that in order to find order in a quasi-isolated system we exploit reproducibility and perform many observations and experiments on many sufficiently similar systems and on the same system itself. Thus the possibility of obtaining data that can be meaningfully compared and from which order can be extracted depends on the existence of many systems other than the one being investigated and on the continued existence in time of the system itself that is under investigation. The other systems serve as reference with which the behavior of the investigated system can be compared, and the continued existence in time of the investigated system allows its behavior to be compared with itself.

That reminds us of the Mach principle (section *Quasi-isolated System and Surroundings* of the preceding chapter), that the origin of inertia of a body lies with all the other bodies in the universe. There, too, the other bodies are needed to serve as reference for comparing the body's behavior, reference for its uniform straight-line or other motion. And there, too, the body's continued existence in time is needed to compare its behavior, its motion in the presence and in the absence of forces, with itself. Just as we know

Laws of Nature (Continued)

of no way to affect the inertia of bodies, so do we not know of any way to change the laws of nature for quasi-isolated systems. All that leads us to enunciate a Mach-like principle, which I called the *extended Mach principle* and which is a generalization of the Mach principle. It is that the origin of the laws of nature for quasi-isolated systems lies with the totality of all systems, i.e., with the universe as a whole. Or more succinctly, the origin of the laws of nature is the universe as a whole. (That includes the standard Mach principle as a special case, since inertia is a lawful aspect of nature, as Newton taught us by his first law.)

Additional support for the extended Mach principle, or at least a hint in its direction, is offered by the following reasoning. Let us assume that in some sense it is meaningful to consider the laws of nature for quasi-isolated systems as possessing an origin, and let us inquire into the location of that origin. Could it be within each quasi-isolated system itself? That is not reasonable, since the same laws of nature are found for different systems. Could the origin of the laws of nature be the immediate surroundings of each quasi-isolated system, say within the volume of a room or so? That is ruled out, because the same laws of nature are found in very different surroundings. Could the origin be our local region of space, say the solar system or our galaxy, of which the solar system is a constituent? That is not reasonable either, since astronomical observations seem to indicate, or at least are consistent with the assumption, that the laws of nature are the same everywhere and under very varied conditions. Thus we are drawn to the whole universe as the origin of the laws of nature.

Yet another line of suggestive reasoning is this. We know that the same laws of nature are valid, wherever relevant, for all quasi-isolated systems, or at least, as mentioned in the preceding paragraph, our observations are consistent with that. We might find some indication about the origin of the laws of nature by asking what else quasi-isolated systems have in common (besides their being quasi-isolated, which is a *sine qua non* for their exhibiting laws of nature to begin with). The only answer I can come up with is that they are all part of the same universe, which indeed suggests the extended Mach principle.

(As an aside, tying together various subjects discussed in the preceding chapter, we might note that this answer is that of a realist, while an idealist would say that all those quasi-isolated systems are being observed by *Homo sapiens*. From the standpoint of the hybrid position presented in the section *Realism and Idealism* of the preceding chapter we could say that the origin of the order that is common to all quasi-isolated systems is the whole universe, while the origin of the laws that are formed from this order is we ourselves. But let us not unduly complicate matters with those considerations.)

Just as in the case of the standard Mach principle, the extended Mach principle must not be understood as a theory explaining the laws of nature. It is rather a metaphysical guiding principle, offering us a framework for organizing our science concepts. And as in the case of the standard principle, the extended one might, on the one hand, be viewed reductionistically as indicating the existence of some influence by which the whole universe imposes behavior on small parts of itself. Then we might try to grasp it scientifically by performing experiments to search for and study such an influence. On the other hand, the extended Mach principle might be viewed holistically. Then it would be inappropriate to consider quasi-isolated systems as being meaningfully separable from the rest of the universe, and the behavior of such systems would result unanalyzably from their very condition of being parts of the whole universe.

An interesting point worth mentioning is the following. According to the standard Mach principle, if the rest of the universe were metaphysically and hypothetically taken away (while in some way leaving a very uninfluential observer), a body would have no inertia. That is because there would then be no reference against which its motion could be determined; there would be no way of knowing if its motion were uniform, in a straight line, or both.

Laws of Nature (Continued)

(There would also be no other bodies around to exert forces on it.)

Similarly, according to the extended Mach principle, if the rest of the universe were taken away from a quasi-isolated system, not only would the system have no inertia, but all laws of nature would cease to hold for it. There would then be no reference for its behavior, nothing with which its evolution could be compared. The quasi-isolated system would then be truly isolated and would constitute a whole universe in itself. Science cannot tell us about the fate of such a system, since, as we saw in chapter 4, science is incapable of framing laws of behavior for the universe and, *a fortiori*, for *universes*. If one would like to speculate metaphysically about its fate, one is, of course, welcome to do so. But there is no compelling reason to assume that the laws of nature would continue to be valid for it. The extended Mach principle says they would not. In any case, removing the rest of the universe already belongs to the domain of metaphysics, as do the Mach principle and its extension themselves. And as long as we are involved in metaphysical deliberations anyway, I might append the stinger that there is no compelling reason to think that even space or time would be relevant to such a system.

In summary, then, what do we know of the origin of the laws of nature? Really practically nothing, if by "know" we mean "understand through science." It is not even clear if the concept of origin is scientifically meaningful for the laws of nature at all. But that should not be surprising, since an origin for the laws of nature would appear to be so fundamental a concept that it very well might not be comprehensible by science. (See chapter 2.) So we philosophize, and our metaphysical considerations lead us to the extended Mach principle, that the origin of the laws of nature is the whole universe. The extended Mach principle can serve as a guide in humankind's scientific endeavor, just as the standard Mach principle has served. Einstein was greatly influenced by the Mach principle, when he developed the general theory of relativity. (That fact was mentioned in passing in the section *Quasi-isolated System and Surroundings* of the preceding chapter.)

WHENCE ORDER?

We have finally reached consideration of the apparent contradiction in the orderly behavior of quasi-isolated systems within the inherently orderless universe, in the existence of laws of nature within the capricious cosmos. In fact, the situation has worsened from what it was at the beginning of the chapter. There the contradiction was apparent when the orderly behavior of a part of the universe was viewed as part of the orderless behavior of the whole universe. But along the way from there to here we learned about the extended Mach principle, by which the orderly behavior of a part of the universe is considered to be not only part of the orderless behavior of the whole universe, but actually brought about by it. In other words, what previously looked like a contradiction now appears to be a paradox: How can orderlessness bring about order? How can inherent lawlessness give rise to laws of nature?

Well, to answer one question by another: Why not? If we think about it, there is really no logical imperative that orderlessness bring about only orderlessness, that lawlessness give rise only to lawlessness. Why should orderlessness not have orderly aspects, if only to an approximation (although, of course, it does not have to)? And why should not lawlessness possess lawful aspects, if only approximately (while, again, it does not have to)? Indeed, we are already familiar with such occurrences, although in other situations, and we will consider three such examples. But please do not look for strict analogy with the main object of our investigation. Whereas the orderless universe is considered to bring about orderly behavior of parts of itself, in each of our examples the orderly behavior of a compound system is brought about by the combined orderless behaviors of its parts. The point being ex-

Laws of Nature (Continued)

emplified is that order can arise form orderlessness and law from lawlessness.

First example: Consider the human population of some country. Each individual behaves in a way that in many respects is far from orderly. He or she makes or loses money in an unpredictable manner; falls sick, recovers, and dies quite randomly; drives a vehicle erratically; and produces babies or refrains from producing them by whim and chance. Yet the experts who are involved with such matters can predict with some degree of accuracy for the *whole* population such things as birthrate, death rate, average hospital occupancy, savings rate, traffic accident rate, annual total of workdays lost due to illness, etc.

Second example: Consider a gas in a closed container, such as air in an otherwise empty corked wine bottle. Because of the immense number of gas molecules in the container under ordinary conditions (roughly of the order of a thousand million million million), the behavior of each individual molecule as it flies around, colliding with other molecules and bouncing off the walls of the container, is unpredictable in practice. Still, properties of the *whole* gas that result from the combined behaviors of its molecules, such as its temperature, its volume, and its pressure on the container walls, do obey orderly relations. For example, if the gas is compressed to half its volume while its temperature is kept constant, its pressure will double.

Third example: A radioactive isotope, such as C^{14} (carbon-14, an isotope of carbon, an important use of which is the determination of the age of archeological finds), is an isotope (a form of a chemical element) the nuclei of whose atoms are unstable and tend to "decay" (to use conventional jargon) by rearranging their internal structure while ejecting one or more subnuclear particles. It turns out to be impossible to predict when any individual nucleus will decay. (Quantum theory tells us that is a matter of principle, not due to our technical inadequacy.)

Nevertheless, a whole sample of a radioactive isotope, containing a very large number of atoms, obeys a very simple law with respect to decay: During equal time intervals equal fractions (percentages) of the original sample will decay. Thus, if half the sample decays, say, in a year, then during the next year half of the undecayed sample, i.e., half of the remaining half, or a quarter of the original sample, will decay, leaving a quarter of the original sample. During the third year a half of that quarter, or an eighth, of the original sample will decay, leaving an eighth. And so on. (By the way, the *half-life* of a radioactive isotope is, as in our example, the time required for the decay of half of any sample of the isotope.)

So we see how orderlessness can bring about order. In the examples that happens by means of an averaging out of the orderlessness of the constituents, resulting in certain orderly behavior of the total system. For the universe as a whole, however, it is the total that is orderless and yet is supposed to bring about orderly behavior of parts of itself. I would be overjoyed if only I could present a good example of that. But I have not yet found one and suggestions are welcome.

The idea of order from orderlessness and law from lawlessness, then, is not so strange after all. And what we have done in this chapter is to use the sneaky rhetorical and didactic trick of setting up a straw man only to knock it down. Since we do find in nature order, predictability, and law (for quasi-isolated systems), it follows that the inherently orderless, unpredictable, lawless cosmos indeed possesses orderly, predictable, and lawful aspects. That need not be precisely true, however, but must at least be valid to a sufficiently good approximation over a time span containing the period of time humans have been making observations and for a region of space containing us and the range of our astronomical observations. Whether nature holds surprises for us over longer time spans or over larger spatial regions is a great unknown. It certainly is possible in principle that the order, predictability, and laws we have discovered will be found to break down in the future or will be found invalid in newly investigated regions of space as

Laws of Nature (Continued)

we extend our observational reach. But such speculations are beyond science. (See chapter 3.)

So the order, predictability, and laws we find in nature are among the (approximately) orderly, predictable, and lawful aspects of the behavior of the universe. What about the orderless, unpredictable, and lawless aspects of the behavior of the universe? Do we find any of that in nature? Does nature have its "dark" sides, phenomena that appear to lie outside the framework of science? Indeed it does! An example on the astronomical scale is that, as the range of astronomical telescopes has been increasing, it has never been found possible, and scientists do not feel it will ever be possible, to predict just where in space the next galaxy, galaxy cluster, or supercluster will be discovered. In other words, the locations of galaxies and their clusters and superclusters in space appear to be an orderless, unpredictable, lawless aspect of nature. Truly cosmic-scale unpredictability, however, is undetectable by us, because the time spans characteristic of the evolution of the universe are immensely longer than the length of time humans have been making scientific observations.

On the scale of the very small we have an extremely important example of unpredictability in nature: It appears to be impossible to predict individual submicroscopic events. That was mentioned earlier in the third example of order from orderlessness in connection with the decay of nuclei of radioactive isotopes. In fact, that unpredictability seems to hold for submicroscopic events of all kinds, such as the absorption and emission of light by atoms and molecules, the chemical bonding and dissociation of atoms and molecules, nuclear fission (the decomposition of a heavy nucleus into lighter nuclei), and nuclear fusion (the merging of light nuclei to form a heavier one). According to quantum theory, as mentioned earlier, that unpredictability is a matter of principle and not merely a technical obstacle eventually to be overcome.

On the human scale the situation is not clear. Some claim the occurrence of phenomena that might be manifestations of the orderlessness of the universe as a whole. I am willing to be open-minded about that, but I am not aware of ever having experienced such a phenomenon myself. If those claims are valid and if at least some of the kinds of reported phenomena persist yet remain as irreproducible and unpredictable as they have so far proved to be, it would seem that we do have human-scale orderlessness and lawlessness in nature. Those phenomena go under names like anomalous events, transient phenomena, miracles, and parapsychology, where the latter includes effects such as extrasensory perception (ESP), telepathy, telekinesis, and clairvoyance. We discussed them briefly in the section *Capricious Cosmos* of chapter 4.

Of course, it should not surprise us that nature possesses an unpredictable side. That is only to be expected, since the universe as a whole is inherently unpredictable. Really, it might be considered remarkable that nature possesses any *predictable* side at all. We have seen that it does, since we do discover laws of nature. And we have seen that there is nothing paradoxical or contradictory in that. But *why* that is so is a valid question. Why does the capricious cosmos have (approximately) orderly aspects? That is a very deep question. Here we are not looking for an explanation of merely this law of nature or that (see chapter 2). Here we are looking for an explanation of the very existence of laws of nature. Why is there order in nature?

Now, a satisfying scientific explanation, i.e., a satisfying theory, of the existence of the laws of nature should entail something even more fundamental than the existence of the laws of nature and also even more fundamental than the universe as a whole, which, by the extended Mach principle (see the previous section), is considered to give rise to the laws of nature. (See the section *Generality and Fundamentality* of chapter 2.) Well, that is a tall order! The only natural thing I can think of that is conventionally more fundamental than the existence of the laws of nature is the universe as a whole. And nothing at all is conventionally more fundamental than the universe as a whole, since it is, by its very definition, all of nature. Thus we find ourselves at the end of science's explanatory power: The existence of laws of nature within the capricious cosmos appears to be unexplainable by anything that is conventionally more fundamental than itself.

Yet the concept of fundamentality lies outside the strict limits of nature and belongs to the domain of metaphysics, as mentioned in the section *Science and Metaphysics* of chapter 3. Nature does not impose that concept on us, and it legitimately depends on one's world view. There exists an unconventional approach within science called the anthropic principle, which holds that in a certain sense the existence of human beings is more fundamental than the universe as a whole. (That was mentioned in the section *Generality and Fundamentality* of chapter 2.) That point of view is just as valid as the conventional one, that there is nothing more fundamental than the universe as a whole. Yet the anthropic principle allows a scientific explanation of the existence of order within the orderless universe, while the conventional viewpoint of fundamentality, as we just saw, does not. That explanation is worked out in detail in the following chapter.

SUMMARY

The Mach principle, that the origin of inertia lies with all the matter in the universe, is generalized to the extended Mach principle, that the origin of the laws of nature for quasi-isolated systems lies with the universe as a whole. Thus the inherently orderless universe must possess (approximately) orderly, lawful aspects, which are the order, predictability, and laws we find in nature. We also find evidence of nature's orderlessness on various scales, including the submicroscopic scale, as described by quantum theory. Science can offer no conventional explanation of the existence of laws of nature within the capricious cosmos.

BIBLIOGRAPHY

For the Mach principle, see: chapter 13 of R. K. Adair, *The Great Design: Particles, Fields, and Creation* (Oxford: Oxford University Press, 1987); and chapter 3 of J. Silk, *The Big Bang*, revised and updated ed. (San Francisco: Freeman, 1989).

For laws of nature, see: chapters 5 and 6 of J. D. Barrow, *The World Within the World* (Oxford: Oxford University Press, 1988); chapter 6 of J. D. Barrow and J. Silk, *The Left Hand of Creation: The Origin and Evolution of the Expanding Universe* (London: Heinemann, 1983); chapters 1 and 2 of R. P. Feynman, *The Character of Physical Law* (Cambridge, Mass.: MIT Press, 1965); part 3 of H. R. Pagels, *The Cosmic Code: Quantum Physics as the Language of Nature* (New York: Simon and Schuster, 1982); and chapter 19 of D. Park, *The How and the Why* (Princeton, N.J.: Princeton University Press, 1988).

Earlier in this chapter—in the section *Whence Order?*—I stated that I would be overjoyed to have a good example of the orderless universe bringing about orderly behavior of parts of itself. Well, I still do not have one. But there is a phenomenon that just might serve as an example, although not an example that I am very happy about. The phenomenon is that of self-organizing systems. What happens is that certain systems can spontaneously generate and maintain local order within a generally chaotic environment. For more details, see: I. Prigogine and I. Stengers, *Order Out of Chaos: Man's New Dialogue with Nature* (New York: Bantam Books, 1984); and chapter 6 of Barrow, above.

For quantum unpredictability, see: chapter 10 of Adair, above; chapter 3 of Barrow, above; chapter 8 of P. C. W. Davies, *God and the New Physics* (New York: Simon and Schuster, 1983); chapters 1–3 of P. C. W. Davies, *Other Worlds: A Portrait of Nature in Rebellion: Space, Superspace and the Quantum Universe* (New York: Simon and Schuster, 1980); chapter 1 of P. C. W. Davies and J. R. Brown, eds., *The Ghost in the Atom: A Discussion of the*

Laws of Nature (Continued)

Mysteries of Quantum Physics (Cambridge: Cambridge University Press, 1986); chapter 6 of Feynman, above; chapters 3 and 4 of H. Fritzsch, *The Creation of Matter: The Universe from Beginning to End* (New York: Basic Books, 1984); chapter 4 of S. W. Hawking, *A Brief History of Time: From the Big Bang to Black Holes* (London: Bantam, 1988); chapter 1 of T. Hey and P. Walters, *The Quantum Universe* (Cambridge: Cambridge University Press, 1987); chapter 6 of part 1 of Pagels, above; chapters 15 and 16 of Park, above; chapters 1–6 of J. C. Polkinghorne, *The Quantum World* (Princeton, N.J.: Princeton University Press, 1984); chapter 1 of A. Rae, *Quantum Physics: Illusion or Reality* (Cambridge: Cambridge University Press, 1986); and chapter 10 of F. Rohrlich, *From Paradox to Reality: Our New Concepts of the Physical World* (Cambridge: Cambridge University Press, 1987).

7 We and the Universe

Space—dimension of our being.
Time—dimension of becoming.
What of space without our being?
Time—without becoming? Or with
No awareness of becoming?
All for us? Because of us?
Explained by us by our becoming?
Simple explanation that.
Still, though it well might warm the soul,
Can never satisfy mind's quest.

HUMAN SCIENCE

Recall the definition of science as *our* attempt to understand the reproducible and predictable aspects of nature, where I deliberately emphasize "our." I stated in the section *Definition* of chapter 1 that the seemingly innocuous qualifier "our" actually carries a heavy load of implication. It tells us that the source of science is within ourselves, that science, although having to do with nature, is actually a human endeavor. Nature, presumably, would go its merry way whether we were around or not and whether we tried to understand it or not. But without our curiosity and urge to understand, *science* would not exist.

From the position of idealism (see the section *Realism and Idealism* of chapter 5) the essential involvement of science with human beings is obvious and merely a matter of definition, since the reproducibility, order, predictability, etc., we find in nature are taken, according to that position, to be mental constructs. But even from the viewpoint of extreme realism it should be clear that science is a human endeavor. In order to do science we must form a conception of nature. And it must be admitted that, as real as realists would have nature and its reproducibility, order, and so on be, our conception of all that is formed only after filtering by the interactions between us and the rest of the universe (observation and measurement), filtering by our senses, processing by our nervous system, conscious awareness of our perceptions, and processing of our perceptions by our consciousness. Then we try to understand what we conceive, and we consider it understood when we have found explanations that satisfy us.

Dolphins, for example, probably have some kind of dolphin science, which would be their attempt to understand nature as they conceive it. Their conception of the universe is formed out of their perceptions, which are obtained through their senses and nervous system, and processed by their consciousness. Dolphins' conception of the universe is in all likelihood quite different from ours, since their senses are known to be different from ours and their consciousness is certainly so too. Their science would involve the devising of explanations that are satisfying to them. It is hard to know what kinds of explanations are satisfying to dolphins, but in sum we can reasonably assume that their science is very different from ours. Yet dolphin science is just as valid for them as ours is for us.

Thus *Homo sapiens* plays an essential role in (human) science in that: (1) science is a by-product of *our* existence; (2) it is *our* conception of nature that we are attempting to explain; and (3) a valid explanation is one that satisfies *us*. Point 1 is obvious. Point 2 is nearly obvious: Conception of nature can well be assumed to be species-dependent, and it is surely not dolphins' or dogs' conception that we are attempting to explain. (By the way, even among people conception of nature is known to be culture-dependent. Thus science should be, and in fact is, different in different cultures as well as in the same culture at different times, as the culture evolves. To the extent we are approaching something like a world culture, at least with regard to conception of nature, we can speak of a "universal" science. But we leave that issue aside.) Point 3 is also nearly obvious, especially after our extensive discussion in chapter 2: Nature imposes no criteria of acceptability for theories; a theory is acceptable when it satisfies *our* feeling that something is indeed being explained.

ANTHROPIC PRINCIPLE

In addition to our essential role in science, i.e., in our attempt to understand nature, we also have a role in the goings-on of nature itself, since we humans are definitely part of the material universe. Our existence being a natural phenomenon, our curiosity drives us to try to understand it within the framework of science, to explain it by other aspects of nature. Charles Robert Darwin's (1809–82) theory of biological evolution and its modern versions are such attempts. (Holders of transcendent world views might try to explain our existence outside the framework of science. Creationism is an example of such an attempt.) Moreover, to invert matters, the existence of human beings, as a natural phenomenon, can in principle serve as an explanation for other natural phenomena. For example, our existence explains very well much of the environmental pollution on this planet. And our existence might possibly explain other, more general and fundamental aspects of nature. The *anthropic principle* is, for a start, that the existence of

Homo sapiens may, within the framework of science, serve as an explanation for phenomena and aspects of nature. (In the following we will make an important addition to that statement.)

At this point it might be useful to reiterate: We are taking the term "understand" to mean "be able to explain." By "explain" we mean "give reasons for." An explanation is a reason (or a set of reasons) for something. (See the section *Definition* of chapter 1.) A theory is a scientific explanation (see chapter 2).

Consider the following. The astronomical data, as interpreted according to the cosmological schemes currently in vogue, seem to indicate that the age of the universe is about 15 thousand million years. Why is the age of the universe just that and not otherwise? If it were much less, by the same cosmological schemes, there would not have been sufficient time for the production (in stars) of the heavier chemical elements necessary for our existence. And if it were much more, all the stars would have burned out and we could not survive. Therefore we should not be surprised to find the age of the universe that we do. Note how the question about the age of the universe is answered in terms of why we cannot be existing and observing it in any era but the present one. Our existence is used to explain, that is, to give a reason for, the age of the universe we discover through our observations. That is an application of the anthropic principle. Admittedly, the age of the universe, which is what is being explained here, does seem to be a rather incidental aspect of nature, since it is changing all the time. For an anthropic explanation of something apparently more fundamental, consider the following.

The anthropic principle has been invoked, among other applications, to explain the strength of the gravitational force, for which no conventional, i.e., nonanthropic, explanation is presently known. Without going into technical details, it seems that if the strength of the gravitational force were much different from what it is, stars would evolve differently from the way they did and do, and in such a manner to preclude the formation of planets and presumably also the development of planet-dwelling observers such as ourselves. Hence our existence implies that the gravitational force has the strength it does and none other; our existence is a logically sufficient condition for the gravitational force to have the strength it does. That is a valid reason, and thus our existence can be taken as explaining the strength of the gravitational force. A similar explanation can be made also for the strength of the nuclear force binding protons and neutrons together to form atomic nuclei.

We are doing some legitimate logical manipulation here. If you are unfamiliar with it or a bit rusty, let us briefly review the idea. Our reasoning is based on the following: If the nonexistence of B implies the nonexistence of A, then, and completely equivalently, the existence of A implies the existence of B. For example, "No clouds in the sky implies no rain" is equivalent to "Rain implies clouds in the sky." In the above anthropic explanation. "The gravitational force not having its observed strength implies our nonexistence" is equivalent to "Our existence implies the observed strength for the gravitational force."

And some terminology: When the existence of A implies the existence of B, then the former is termed a *sufficient condition* for the latter, while the latter is called a *necessary condition* for the former, and the latter can be said to *follow from* the former. For example, rain is a sufficient condition for clouds in the sky, clouds in the sky are a necessary condition for rain, and there being clouds in the sky follows from there being rain. Also: When the existence of A implies the existence of B, then the former is a valid reason for the latter and thus explains the latter. Conversely and equivalently, as we have seen, the nonexistence of B is then a valid reason for the nonexistence of A and explains it. For our rain example, rain can be considered a reason and an explanation for there being clouds in the sky. (Whether such reason and explanation are satisfying is another matter.) Then conversely and equivalently, no clouds in the sky is a reason and explanation for no rain.

Let us use the anthropic explanation of the strength of the gravitational force as a case study to help clarify the anthropic principle.

The explanation runs into two difficulties, which we might call, respectively, subjective and objective. The subjective difficulty is that scientists, and most likely you too, just do not *feel* that any explaining is being done. That is the standard objection to the anthropic principle. Referring to chapter 2, recall that we want that which is explaining to logically imply that which is being explained, to be more general, more fundamental, more unifying, and simpler than that which is being explained, and we would also like the former to be the cause of the latter.

Of all those desirable attributes it seems that only the first is true, the *sine qua non* of a scientific explanation, without which the explanation would not have been proposed to begin with. As we saw, our existence indeed logically implies the actual strength of the gravitational force, because, if it were much different, we would not exist. On the other hand, our existence does not at all seem to be more general, more fundamental, more unifying, or simpler than the strength of the gravitational force. Neither do we perceive our existence as causing the latter. If anything, the opposite would appear to hold—that the strength of the gravitational force is more general, more fundamental, more unifying, and simpler than our existence, and even possibly that the strength of the gravitational force is part of the cause of our existence.

Indeed, we must gracefully concede lack of generality, unification, simplicity, and causation in that anthropic explanation. And that is the subjective difficulty of the anthropic principle when it is applied to the explanation of apparently rather fundamental aspects of nature. Without those it is hard to feel that an anthropic explanation is an acceptable explanation. We must not concede lack of fundamentality, however. After all, who are we to tell the universe what is more or less fundamental for it? Viewed holistically, the universe is an integrated, all-encompassing whole, of which all aspects and phenomena are interrelated and interdependent. We cannot experiment with variations of the universe, so we cannot know which of its aspects and phenomena, if any, are more or less fundamental. Thus absolute fundamentality is an empty concept in science. The most we can do is point out what aspects of nature *seem* more or less fundamental *to us* on the basis of *our* understanding of nature.

Fundamentality, therefore, is both relative to our state of understanding and a matter of convention. Having recognized that fact, we can come to the realization that our existence (that of each of us generalized to *Homo sapiens*) is the most fundamental aspect of nature *for us*, as it underlies our very perception of other aspects of nature. Some people might not consider that reasoning to be very satisfactory. But I think it is quite valid, when the nature of science, as an intrinsically *human* endeavor, is taken into account. After all, in the whole of nature what phenomenon are we most sure of, have least doubts about, have the most confidence in, if not *our own existence*? Consider what a person deprived of all sensory perception would be aware of. Whereas it is fairly innocuous to doubt any other natural phenomenon, consider the paradox involved in doubting one's own existence. Indeed, *cogito ergo sum* (I think, therefore I am)!

So, although we concede a subjective difficulty for the anthropic explanation in the case under study and in general, the point of fundamentality is not part of that difficulty, when considered from the enlightened vantage we just gained. Thus the anthropic principle not only states that the existence of *Homo sapiens* may serve as an explanation for other aspects of nature, but the principle can be viewed as offering the most fundamental explanations, since it bases them on the most fundamental natural phenomenon we have, our own existence. We can now state the anthropic principle in its full form: The existence of *Homo sapiens* may, within the framework of science, serve as an explanation for phenomena and aspects of nature, and moreover, such explanations are the most fundamental.

The objective difficulty of the anthropic explanation in our case study, and in other similar cases, is what I call the invariant context problem. As we are aware, for an "explanation" to be an explanation, at the very least that which is explaining must logically imply that which is being explained. In our case the existence of *Homo sapiens* must logically imply that the strength of the gravitational force is as it actually is, or, the actual strength of the gravitational force must follow logically from our existence, or, if the strength were different we could not exist. (Refer to our logic review earlier in this section.) Now it has indeed been shown that, if the strength of the gravitational force were much different from what it actually is, we would not be around to wonder about it. Thus the actual strength does seem to follow from the existence of *Homo sapiens*.

But it is important to be aware of the context of that explanation. While we are considering the hypothetical possibility of varying the strength of the gravitational force, we are tacitly assuming that no other aspect of the universe, no other law of nature, nothing else at all, is being varied. That is the invariant context problem, because we can never be sure that a change in the strength of the gravitational force cannot be compensated for by some concomitant change in other laws of nature. That kind of qualification must be kept constantly in mind in all applications of the anthropic principle, even if not stated explicitly. Almost all applications of the anthropic principle assume an invariant context, which qualifies their explanatory power. We can almost never be sure that some concomitant variation in the context will not compensate for the assumed variation in the situation being explained, so that the existence of *Homo sapiens* will not be precluded after all.

So where do we stand with the anthropic principle? Due to the subjective difficulty presented earlier, the anthropic principle should be resorted to only when no conventional explanation is available, i.e., only for explaining aspects of nature that appear to be so fundamental that we are hard put or at a loss to find an aspect of nature that conventionally appears to be even more fundamental. Such an aspect, for example, might be the laws of nature, the nonseparability of nature, space, or time. When we run into such a situation, we can declare, "Aha! We thought that aspect is so fundamental. But we are now anthropically enlightened, and we

We and the Universe

know that our existence is even more fundamental (*cogito ergo sum!*)." And we might devise an anthropic explanation for so fundamental an aspect of nature.

True, it would not be a very satisfying explanation, since it would not meet the criteria of generality, unification, simplicity, or causation. But it would fulfill the criteria of logical implication and fundamentality, and would thus be a viable explanation, where the conventional alternative offers no explanation at all. With anthropic enlightenment and good will one can get used to anthropic explanations, especially when no conventional ones are available. Nevertheless, the explanations offered by the anthropic principle are liable to be qualified by the assumption of invariant context, which is their objective difficulty.

Explanatory chains can be set up in nature. Say we wish to explain some aspect or phenomenon of nature. We look for another, conventionally more fundamental aspect or phenomenon of nature that explains the former. Upon finding it we are back to square one, because we are not satisfied leaving this aspect unexplained either. And so it goes, from aspect to conventionally more fundamental aspect. Until we have gotten so fundamental that we are hard put or at a loss to find any conventionally more fundamental aspect to explain our last explaining aspect, except perhaps the universe itself, which amounts to no explanation at all. At that stage we recognize that our existence is really the most fundamental aspect of nature *for us* and perhaps succeed in thus explaining the last extremely fundamental aspect we were trying to explain.

Let us say we want to explain the existence of *Homo sapiens*, which conventionally appears to be quite an incidental phenomenon of nature. We might appeal to Darwinian evolution at the first level of explanation. Then we might explain Darwinian evolution by means of molecular biology at the next level. The following level might be an explanation of molecular biology through biochemistry. Then we might explain the latter by physics, i.e., by quantum theory and the other basic laws of nature. And then what? We have no conventional explanation for quantum theory and the other basic laws of nature. So rather than end the explanatory chain at that point, we could invoke the anthropic principle and perhaps explain quantum theory and the other basic laws of nature by the existence of *Homo sapiens* as the most fundamental aspect of nature.

What a beautiful circularity: our existence (viewed fundamentally) explaining our existence (viewed incidentally)! Should that bother us? Not at all. We realize that we are not dealing with just any universe or with some general model of a universe, but rather with the only universe we have. And we realize that that universe *contains us in an essential way* by the very fact of *our* investigating it. A relevant concept here is "self-reference": The object of the investigation intrinsically contains its investigators as part of itself; the investigators are investigating an object of which they are intrinsically a part. That is the holistic world view within which the anthropic principle operates.

Let us make a clarification at this point. It should be emphasized that the anthropic principle does not imply causation, and an anthropic explanation does not mean that the existence of *Homo sapiens* is the cause of whatever it is explaining. Indeed, that is one of the unsatisfactory properties of anthropic explanations. There is no teleology (no design or purpose) or religion here, and the anthropic principle is no return to anthropocentrism, i.e., it does not reinstate humanity to its historically former privileged position as the center of nature. It does, however, recognize our privileged position as the center of science (one of the centers, anyway), since science is, it cannot be overemphasized, a *human* endeavor. Neither is it claimed that *Homo sapiens* is the reason for the universe. That would be metaphysics, while the anthropic principle is a scientific principle. An anthropic explanation is, however, an explanation of last resort, to be used only when no conventional explanation is available, in order to provide, where applicable, *some* (however unsatisfactory) scientific explanation rather than having none at all.

Some more comments are in order here. As we derived it above,

We and the Universe

the anthropic principle is: The existence of *Homo sapiens* may, within the framework of science, serve as an explanation for phenomena and aspects of nature, and moreover, such explanations are the most fundamental. In order to give a more nearly balanced perspective, I should mention that not all scientists accept the anthropic principle as stated here; some object to the second part and some to the whole thing. In addition, other versions of the anthropic principle have been proposed, and they too have their objectors. Actually the term "anthropic principle" covers a variety of principles, all having the common feature of somehow involving the existence of *Homo sapiens*. Support for or objection to any one version is not necessarily valid for or against other versions.

I will not burden you with other versions. But I will add that the anthropic principle we derived above is the only version I know of that both is a *scientific* principle (some versions are metaphysical) and, in my opinion, possesses deep significance. It not only is concerned with what is going on in nature but is also concerned with science, our attempt to understand what is going on in nature, as a *human* endeavor. It recognizes that we, as investigators of nature, are intrinsically part of the object of our investigation, so that in the final reckoning we cannot avoid coming back to ourselves in some manner or other. We can thus appreciate Arthur Stanley Eddington's (1882–1944) beautiful description, though let us not interpret it too extremely: "We have found a strange footprint on the shores of the unknown. We have devised profound theories, one after another, to account for its origin. At last, we have succeeded in reconstructing the creature that made the footprint. And lo! it is our own."

WHENCE ORDER? (AGAIN)

In the section *Whence Order?* of the preceding chapter we raised the question of why there is order within the capricious cosmos, why, indeed, nature possesses any predictable side at all. There we saw that no conventional answer to that question should be expected, since what we want to explain appears to be too fundamental for conventional explanation. Thus the situation is ripe for an anthropic explanation, if it can be shown that some predictability and order are necessary for our existence.

Well, we cannot give a rigorous proof, but we can give a plausibility argument, which might run as follows: Without predictability we could not reliably control our bodies or anything else and could not survive our interactions with our unpredictable environment and with each other. Also, consider the degree of constancy of the laws of nature, both throughout space and over time, that is necessary for our survival as individuals and as the human race, for us to have stable, reliable memories as individuals and as a society, and so on. Due to our physical complexity it is very plausible that the bounds on the possible variability of the laws of nature are extremely tight, at least throughout some sufficiently large volume of space and over some sufficiently long period of time.

If we take that plausibility argument seriously, the anthropic reasoning (refer to the logic review in the previous section) then runs: Since lack of order implies our nonexistence, it is logically equivalent that our existence implies order. Thus, since we do exist, the capricious cosmos must have orderly aspects, and those orderly aspects must and do affect us strongly enough to allow our existence. They are the laws of nature we discover. The orderless aspects of the universe must and do affect us only weakly; otherwise, by the same reasoning, we could not survive.

As we saw in the section *Extended Mach Principle* of the preceding chapter, we know of submicroscopic orderlessness, the essential quantum unpredictability on the scale of the very small. It affects us by causing mutations, maybe cancer, and perhaps other unpleasant results. But although individual personal tragedies may be grave, the effect on the whole is weak, since most people manage to survive and the human race thrives. (Mutations are an essential component of biological evolution.) Possible human-scale irreproducibility, if it exists, might show up as "anomalous events," "transient phenomena," miracles, parapsychology, etc., which appear only rarely and affect us only slightly on the whole. Such effects were discussed briefly in the section *Capricious Cosmos* of chapter 4. Larger-scale orderlessness does not affect us at all.

So that is how the anthropic principle explains the existence of laws of nature within the capricious cosmos. In that case the explanation is especially unsatisfying, because we cannot even show that our existence implies what it is supposed to explain and so we have to make do with a mere plausibility argument. On the other hand, the anthropic principle does give us some explanation for what would conventionally remain unexplained.

SPACE AND TIME

The conventionally most fundamental aspects of nature are, it seems to me, space, time, and nature's quantum aspect (including submicroscopic unpredictability and uncontrollable correlations). Since there can be nothing conventionally more fundamental than most fundamental, no conventional explanation of space, time, and the quantum is to be expected (although it might be possible to explain one or two of them by the other two or one). So again we have a situation that is ripe for the anthropic principle, where only the existence of *Homo sapiens* (considered fundamentally) might explain what cannot be explained conventionally. And here the anthropic principle can offer a clean, unqualified explanation. We apply the anthropic principle to the explanation of space and time.

To begin with we need working definitions for space and time. As one might well imagine, much has been said and written about that, mostly philosophical. My own preference is: Space is the dimension of being; time is the dimension of becoming. Those definitions are extremely concise, which is an advantage. But their conciseness tends to hide their complexity, which is the complexity inherent to the concepts of dimension, being, and becoming. The concepts of being and becoming have been the subjects of philosophical deliberations for ages, and those deliberations are still going on just as vigorously as ever. For the purpose of the present discussion we will exempt ourselves from entering into those deliberations and will leave the concepts of being and becoming to be understood by each reader in whatever way they are ordinarily understood.

The concept of dimension, however, is more technical and is certainly less familiar to the general public. The term has several related meanings, but the one that is relevant to our discussion is: Dimension is the possibility of assigning a measure to something. Thus we can speak of a musical tone as possessing the dimension of pitch, measured by frequency (in hertz, Hz), the dimension of intensity, measured by power per unit area (in watts per square meter, W/m^2), and the dimension of timbre ("tone quality"), which can be measured also. Or, color is said to possess the dimensions of hue, measured by two "chromaticity coordinates" (numbers between zero and one), of saturation, measured by purity (again a number between zero and one), and of intensity, measured by power per unit area (in watts per square meter, W/m^2). Often a dimension is assigned to each number involved in a measure. Thus color is said to possess four dimensions: two dimensions of hue (two chromaticity coordinates), one dimension of saturation (purity, a single number), and one of intensity (the number of units of power per unit area). The shape of a packing box has three dimensions: its length, its width, and its depth, where each is measured by linear extent in centimeters, say, or in inches.

The ordinary concept of material being always allows the assignment of a measure by the answer to the question Where? For example, Where is it? Moreover, that is the only measure that can always be assigned to it. Thus, by the definition of space as the dimension of being, space is just that possibility of answering the question Where? with regard to being. Our experience teaches us that space is actually three-dimensional, since the question Where?

always requires three numbers for a full reply, for instance, latitude, longitude, and altitude with respect to the Earth.

Becoming involves change. It presupposes being, but is more than mere being in that it always allows the assignment of a measure by the answer to the question When? When did (or does or will) the change occur? for example. So, by the definition of time as the dimension of becoming, time is just that possibility of answering the question When? with regard to becoming. From our experience we know that time is a single dimension, since the question When? can always be answered by a single number, composed, for example, of the date and the "time" (in the sense of clock reading).

Now, we want to use the existence of *Homo sapiens* to explain space and time. The first step is to show that the existence of *Homo sapiens* implies being and becoming. Well, we obviously *are* and we clearly *become*. But to sharpen things a bit let us extract from "the existence of *Homo sapiens*" just "the existence of a learning system." We are systems that exist and learn. Then: Existing is being, and learning implies becoming (specifically, becoming more learned). Thus the existence of *Homo sapiens* implies being and becoming.

The next step of the anthropic explanation of space and time is to recognize that everything implies its own dimension(s). For instance, color implies its dimensions: hue, saturation, and intensity. A rectangle implies its dimensions: length and width. And similarly being implies its dimension, space. And becoming implies its dimension, time.

The third step is to combine the first two steps. The existence of *Homo sapiens* implies being, which in turn implies space. Thus our existence implies space. And our existence also implies becoming, which in turn implies time. Thus our existence also implies time.

The final step of the anthropic explanation is to take the logical implication as a reason and thus as an explanation. Hence the existence of *Homo sapiens* explains space and time. It is as simple as that. A clean, unqualified application of the anthropic principle.

SUMMARY

Science is a human endeavor, since *Homo sapiens* plays an essential role in it in that: (1) science is a by-product of *our* existence; (2) it is *our* conception of nature that we are attempting to explain; and (3) a valid explanation is one that satisfies *us*. The anthropic principle is that the existence of *Homo sapiens* may, within the framework of science, serve as an explanation for phenomena and aspects of nature, and moreover, such explanations are the most fundamental. Fundamentality follows from the fact that our existence is the most fundamental aspect of nature *for us*. Anthropic explanations should be used only as a last resort, for aspects of nature that are apparently so fundamental that no conventional explanation is available. That is because they suffer from the subjective difficulty that our existence seems neither more general, more unifying, nor simpler than whatever it is explaining, and it is not perceived as causing the latter. They also suffer from the objective difficulty of the invariant context problem. The existence of order within the capricious cosmos is given a rather weak anthropic explanation. But a clean, unqualified anthropic explanation is given for space and time.

BIBLIOGRAPHY

For science as a human endeavor, see: B. d'Espagnat, *In Search of Reality* (New York: Springer-Verlag, 1983), which is not too easy reading; E. Harrison, *Masks of the Universe* (New York: Macmillan, 1985); K. Lorenz, *Behind the Mirror: A Search for a Natural History of Human Knowledge* (New York: Harcourt Brace Jovanovich, 1977), which is especially concerned with the evolutionary adaptation of our cognitive facility to the reality of our environment; R. Morris, *Dismantling the Universe: The Nature of Scientific Discovery* (New York: Simon and Schuster, 1983); D. Park, *The How and the Why* (Princeton, N.J.: Princeton University Press,

1988); chapters 1 and 18 of R. K. Adair, *The Great Design: Particles, Fields, and Creation* (Oxford: Oxford University Press, 1987); chapters 1 and 7 of J. D. Barrow, *The World Within the World* (Oxford: Oxford University Press, 1988); part 4 of M. Eigen and R. Winkler, *Laws of the Game: How the Principles of Nature Govern Chance* (New York: Knopf, 1981); chapters 15 and 16 of H. Fritzsch, *The Creation of Matter: The Universe from Beginning to End* (New York: Basic Books, 1984); chapter 9 of J. Monod, *Chance and Necessity: An Essay on the Natural Philosophy of Modern Biology* (New York: Knopf, 1971); part 3 of R. Morris, *The Nature of Reality* (New York: McGraw-Hill, 1987); book 1 of I. Prigogine and I. Stengers, *Order Out of Chaos: Man's New Dialogue with Nature* (New York: Bantam Books, 1984); and chapter 1 of F. Rohrlich, *From Paradox to Reality: Our New Concepts of the Physical World* (Cambridge: Cambridge University Press, 1987).

For discussions, applications, and other versions of the anthropic principle, see: J. D. Barrow and F. J. Tipler, *The Anthropic Cosmological Principle* (Oxford: Oxford University Press, 1986); R. Breuer, *The Anthropic Principle: Man as the Focal Point of Nature* (Boston: Birkhauser, 1991); chapter 18 of Adair, above; chapter 7 of Barrow, above; chapter 5 of J. D. Barrow and J. Silk, *The Left Hand of Creation: The Origin and Evolution of the Expanding Universe* (London: Heinemann, 1983); chapter 5 of P. C. W. Davies, *The Accidental Universe* (Cambridge: Cambridge University Press, 1982); chapter 8 of P. C. W. Davies, *Other Worlds: A Portrait of Nature in Rebellion: Space, Superspace and the Quantum Universe* (New York: Simon and Schuster, 1980); chapter 7 of P. C. W. Davies, *Space and Time in the Modern Universe* (Cambridge: Cambridge University Press, 1977); chapters 1 and 10 of J. Gribbin and M. Rees, *The Stuff of the Universe: Dark Matter, Mankind and the Coincidences of Cosmology* (London: Heinemann, 1989); and chapter 8 of S. W. Hawking, *A Brief History of Time: From the Big Bang to Black Holes* (London: Bantam, 1988).

8 Reality

Glimpsed as through a streaming mist
Or a breeze-blown gauzy veil.
Sensed in ways beyond mind's ken,
So no reason can avail.
But it is! It is! As on we drift
Through the froth of its phenomena.

The nature of reality has been the subject of endless searching philosophical thought, discussion, and argument for ages and ages. In our own age the issue is still far from resolved and continues to arouse undiminished fervor. As science developed from, say, the 1600s to the present, it has taught humankind objective facts about reality that thinkers cannot afford to ignore. And especially in our own century, the implications of the attainments of science, as expressed in relativity and quantum theories, are crucial for our understanding of reality. So much so, that I strongly feel that any educational curriculum worthy of its name must assure some minimal elementary understanding of those theories.

Even so, it must be emphasized that considerations about reality are basically philosophical considerations, even metaphysical, in the sense described in the section *Science and Metaphysics* of chapter 3, since they bear strongly on science. Although science can constrain metaphysical speculations about reality, it cannot totally replace them. Science can go only so far in telling us about reality; reality's ultimate nature is beyond the domain of science. Yet it seems to me that, although it cannot force one to adopt any particular world view, science does strongly hint in a certain direction. And if we are willing to let science guide us, we will indeed find ourselves being led to a particular world view.

METAPHYSICAL POSITIONS

First let us reconsider the metaphysical position of *realism*, which we considered in the section *Realism and Idealism* of chapter 5. There realism was presented as the position that the laws of nature are inherent to the world we observe and are independent of observers. Realism was opposed there to *idealism*, the position that the laws of nature are wholly in the mind of the observer. I proposed a hybrid position, that nature does possess objective order while laws are a human device.

It should now be pointed out that *realism*, in a more fundamental sense of the term, is the metaphysical position that there is an objective, observer-independent underlying reality that we discover and study through our physical senses (and through our measuring and observing instruments as extensions of our senses), that there is a real world "out there" that exists independently of us. The tacit assumption of realism in this sense pervades the presentation in this book. That is because my own world view is basically realist, as in fact is that of almost all scientists. Realism in this sense is usually put in opposition to *positivism*, which is the metaphysical position that only our sense data, derived from measurements and observations, are fundamental. Positivist science is thus expressed in terms of measurements and observations, in terms of phenomena, and refrains from considerations of any underlying reality that would not be directly accessible by our physical senses and their extensions.

Reality

Both realists and positivists agree that our measurements and observations are the source of our scientific knowledge of nature. Positivists claim that it is meaningless to go beyond that, that whatever might be underlying the phenomena, nothing significant can be said about it. Realism goes beyond positivism by assuming that an objective reality underlies natural phenomena and is the ultimate source of the results of our measurements and observations. Thus, according to the realist view, our measurements and observations, as interesting as they might be in themselves, are telling us something about objective underlying reality.

As an example of the difference between realist science and positivist science, consider a hypothetical investigation of a sample of some radioactive material. Say one gram of the material is placed two centimeters away from the window of a Geiger counter. The counter responds with a continual, irregular series of clicks, while the pointer of its meter indicates the average number of clicks per minute, the count rate. (The technical details are not important for our example, but I gave them to make it sound authentic.) Both realist and positivist scientists would be interested in the dependence of the count rate on the distance between the sample and the counter window and its dependence on the thickness and kind of material placed between the sample and the counter window. Both would be interested as well in the dependence of the count rate on time. Both would make many measurements of many kinds, analyze the results, and look for order.

Positivist scientists would not go much beyond that. They would be happy to find interesting order among the measurement results and would be satisfied with that. Realist scientists would also be happy with that, but would not be satisfied. They would look for an underlying reality that brings about the Geiger counter clicks and the dependence of the count rate on distance, intervening material, and time. They would reach an understanding in terms of the sample's consisting of atoms with unstable nuclei, nuclei that spontaneously and unpredictably emit particles. They would picture the particles entering the counter window and causing the counter to click, one click per particle. They would understand that intervening material absorbs those particles, so that less of them reach the counter window. They would understand that the count rate decreases in time because there remain less and less unstable nuclei. They would even suggest new experiments and predict the results in order to confirm their understanding.

It might be interesting to note that positivism carried to the extreme leads to *solipsism*, which, in its extreme, is the metaphysical position that nothing is real except the self. In other words, starting with *cogito ergo sum* (I think, therefore I am), I am assured only of my own existence; the rest is my imagination (including this book and its readers). It must be admitted, it seems to me, that from a purely logical standpoint solipsism is the only compelling metaphysical position among them all. Indeed, the reality to me of everything beyond my own existence is really based on arbitrary assumptions that I choose to make.

For example, it is an assumption—perhaps reasonable, maybe not, but nevertheless arbitrary—that my self is somehow inherently attached to a body. It is a further assumption that my body is equipped with sense organs and that certain sensations I am aware of result from activity of those organs. Then it is also an assumption that the activity of my sense organs is a result of stimulation by some reality external to my body. And so on and so forth. (Am I just imagining that I am hitting imagined keys that bring these imagined words into imagined existence?)

OBJECTIVE REALITY

Let us now proceed step by step and see how science leads us to a particular metaphysical position in regard to reality, which, as can readily be predicted from the preceding paragraphs, is going to be a version of realism. My arguments and justifications will be based on common sense, on our understanding of nature, and on reasonableness, utility, motivation, and conceptual economy (i.e., thrift in using concepts).

To start off, it seems to me that science is urging us to be realists, that we are being encouraged to believe that there is an objective reality, an observer-independent underlying reality that is the same reality for electrons, for galaxies, for tadpoles, and for people. One argument here is based on *motivation*. If one does not believe that there exists an objective reality, then there is no motivation to discover and investigate it. Yet the very existence of science demonstrates that there are many people who are strongly motivated to discover and investigate objective reality. That argument is good for the existence of objective reality, but does not touch upon its objectivity.

However, it is most *reasonable* to believe in its objectivity, that indeed it is observer-independent and is the same underlying reality for neutrons, for stars, for carrots, and for humans. Science goes to much trouble and effort to maximize objectivity by confining itself to the reproducible aspects of nature (see the section *Reproducibility* of chapter 1). Then science finds that nature indeed possesses such aspects, and moreover, that there is no lack of raw material for science to process. And, finally, almost all that raw material is well comprehended by science and is found to fit together beautifully and to mesh into a most elegant conceptual fabric, the whole wonderful world of science. So if it looks, sounds, and smells like objective reality, is it not most *reasonable* and *conceptually economical* to believe that that is just what we have?

Furthermore, it is also *useful* to believe that there is an objective reality, because, motivated to investigate it, we do investigate it and, through technological application, obtain useful results, such as fertilizers, vitamins, telephones, and word processors.

Now that we are convinced to believe in an objective reality, we note that we gain scientific knowledge of it through our physical senses: sight, hearing, smell, taste, touch (pressure), temperature, and others. (Any knowledge of reality gained through other, non-physical channels is of no concern to science, as interesting as it might be otherwise.) Information from our senses is processed by our nervous system, where it interacts with innate and acquired structures, finally emerging in our consciousness. Thus our knowledge of reality involves interaction with the world external to our bodies, filtering through our senses, neural processing, and conscious awareness of the results of all that.

That seems obvious. The very important implication is that scientific knowledge of objective reality is *indirect* knowledge. Direct knowledge would be knowledge that we become aware of without the process just described. It would somehow pass directly from the object of the knowledge into our consciousness. We might call it intuition or belief or feeling. But by its very character science confines us to our *physical* senses in our observations of nature. Here intuition, belief, and feeling are forbidden. "Data" gained by intuiting the reading of a gauge, instead of actually reading it, or by believing that a length is such and such, rather than actually measuring it, or by feeling that a solution has a certain color, instead of actually seeing it, have no part in science. Thus science dooms itself to merely indirect knowledge of the reality it is trying to comprehend.

That fact seems to put off some people, who prefer what they consider to be direct knowledge of reality through nonscientific modes of comprehension over the indirect knowledge of science. Thus they might attempt to know reality by means of meditation, prayer, feeling, intuition, inspiration, awareness, or pure thought, possibly with the help of "awareness-enhancing" or "mind-expanding" drugs. Who but the subject himself or herself can judge whether any knowledge at all is thus gained? And who, including the subject, can judge whether such knowledge has anything whatsoever to do with reality? What can be stated with certainty, though, is that any knowledge thus gained is among the most subjective knowledge imaginable.

Yet scientists too indulge in intuition, inspiration, feeling, and pure thought, as we saw in chapter 2 dealing with theories in science. And not only in their devising and judging of theories do and must scientists act with some degree of irrationality, but intuition, inspiration, and taste can also guide the choice of what aspect of nature to investigate, how to go about it, and what experiments to perform, for example. In fact, it is to a large extent by their superior intuition and inspiration that the greatest scientists achieve their stature in the science community. But at the level of observational data no scientist will ever accept "direct knowledge" in lieu of knowledge gained via the physical senses, however indirect the latter admittedly is.

PERCEIVED REALITY

What it is that we do become aware of via our senses and neural processing might be termed *perceived reality*. It is perceived reality that is the actual subject of science, although we might hope to gain some understanding of objective reality through an understanding of perceived reality. Now since, according to our *understanding of nature*, we are the result of natural evolution, and since we are managing to survive in our ecological niche, it is *reasonable* to believe that perceived reality cannot be much out of tune with objective reality, at least with those aspects of reality that strongly and frequently affect our survival as individuals and as a species. When we perceive food at a certain location at a certain time and strive to get hold of it, we succeed in nourishing ourselves often enough to survive. Or, we survivably often manage to avoid the attacks of tigers (as well as of motor vehicles, their spiritual descendants).

Thus, in the absence of any direct scientific knowledge of objective reality, we should not throw up our hands in despair, but rather should take our perceptions and concepts seriously and let them guide us toward an understanding of objective reality. The argument here is one of *motivation* and *utility*: If we have no hope of approaching objective reality, we might as well give up believing in it. Indeed, it is *reasonable* and *conceptually economical* to believe that our perceptions and concepts even give a *literal* picture of objective reality, at least as long as that belief is tenable.

Even so, we cannot presume that our perceptions and concepts of reality will necessarily give a literal description, or even remain faithful guides, for aspects of reality that do not affect us strongly or frequently, such as for submicroscopic phenomena and for astronomical and larger-scale phenomena. The reason for that is that the argument based on evolutionary adaptation then loses its validity. It does not much matter whether we are adapted to aspects of reality that affect us only weakly or rarely; in any case they do not significantly affect our survival. So when dealing with such aspects of reality, it is *reasonable* to let our belief in our ability to literally describe objective reality through our understanding of perceived reality be contingent on its not leading to undue difficulty. Indeed, we should be prepared to drop our belief in literal description if and whenever it proves to cause more trouble than it is worth.

And it turns out that, when we are dealing with the quantum aspect of nature, our belief in a literal description of objective reality does cause tremendous trouble. Perceived reality at the submicroscopic level of nature, which is well described by quantum theory, is simply unacceptable as a literal description of objective reality. The technicalities of the problem are beyond the scope of this book. They are presented in other books, some of which are listed in the bibliography at the end of this chapter. But the cardinal point can be stated without going into technicality. It is that in quantum theory there is too much dependence on the observer to allow perceived reality to be a literal description of objective reality, which is assumed to be independent of observers.

Let us try to clarify that at least somewhat. Quantum theory is formulated in terms of possible happenings and their probabilities of actually occurring, rather than in strictly deterministic terms of what *will* occur. For example, for some situation quantum theory

Reality

might tell us that a particle can be in a certain range of positions (rather than specifying its exact location). If we make an observation of its position, we will find it at some location within the allowed range. But only then, according to quantum theory, are we allowed to think of the particle as actually being located. Prior to the observation the particle must not be thought of as being located at all. It then possesses only potentiality for location. The act of observation realizes the potentiality and endows the particle with the property of position.

It is not as if prior to the observation the particle had a location that we did not know but which the observation revealed. Quantum theory very clearly rules that out and affirms that it is the act of observation itself that bestows the property of position upon the particle, which previously did not possess that property. That intrinsic dependence of the perceived reality of properties, such as position, on the activities of observers demonstrates that the perceived reality of the quantum aspects of nature cannot be a literal description of an observer-independent, objective reality.

That very well might seem to make little sense, probably even no sense at all. In fact, it might appear altogether downright crazy. Yet that is the way things are. The astounding, mind-boggling quantum aspects of nature are indeed very counterintuitive. But they are no less part of perceived reality for all that. The trouble, of course, is with our limited intuition, which developed in the environment of ordinary-size phenomena that affect us strongly and often. At that scale quantum effects, though valid, are negligible. So we have no intuition for them. Nevertheless, their implications for reality are so crucial that no thinker can afford to ignore them. Thus I repeat my claim at the beginning of this chapter, that no educational curriculum worthy of its name can do without some minimal exposure to quantum theory (as well as relativity theory). If you are not familiar with the ideas of quantum theory, I strongly urge you to make use of the bibliography at the end of the chapter.

PARTIALLY HIDDEN REALITY

Thus we know that even if certain aspects of perceived reality can be assumed to give a literal description of objective reality, there are other aspects that do not. We know that science does not give us full comprehension and understanding of objective reality. Whether science ever will do that is a moot issue at present, but the way things are developing does not offer much hope. So the objective reality science has led us to believe in is partially hidden from us, is a veiled, clouded, fogged reality. Science allows us a few clear glimpses of parts of it as well as provocative hints about more of it. Most of objective reality, however, will very likely remain inaccessible to humankind via science.

It seems to be human nature to want what we do not have and to desire especially what we cannot have. And so with objective reality; the very interesting questions about it are those we cannot presently answer, and the really profound ones will most probably never be answered by science. Questions like: What can be deduced from quantum theory about objective reality? Are space and time, as we perceive them, fundamental aspects of objective reality? (There appear to be indications to the negative.) And if not, how do they tie in with it? To what extent is our perception of nature in terms of localized objects valid for objective reality? (Not very, it seems.) What relation does mind bear to objective reality? And the profoundest questions of all are most likely far beyond human mental ability even to formulate.

Since science does not, and most probably cannot, give us anything near full comprehension and understanding of objective reality, there seems no reason why other, nonscientific modes of comprehension and understanding should not afford us hints and clues to that reality. As long as objective reality, which science guides us to believe in, itself lies beyond the domain of science, let us not shut ourselves off to the possibility that we might possess other channels to it. Indeed, why should not intuition, belief, feeling, art, music, and poetry be allowed to contribute whatever insight they might offer? Such modes of comprehension, as irrational and subjective as they are and perhaps even *because* of their irrationality and subjectivity, might complement, and thus strengthen, the contribution of science to our quest for objective reality.

Let us summarize. Using arguments and justifications based on common sense, on our understanding of nature, and on reasonableness, utility, conceptual economy, and motivation, we have seen how science leads us to the belief in an objective reality. Our scientific knowledge of that reality, however, is indirect. But science should nevertheless give us a literal description of objective reality, at least for those aspects of it that strongly and frequently affect our survival as individuals and as a species. Yet quantum theory cannot be a literal description of that reality. Thus objective reality is partially, very likely mostly, hidden from us. We should not exclude the possibility that other, nonscientific modes of comprehension and understanding might offer insight into objective reality.

TRANSCENDENT REALITY

In earlier chapters we made much of the concept of nature. How does nature fit in with reality? Recall that nature is, for our purpose, the material universe with which we can, or can conceivably, interact, which coincides quite closely with perceived reality. Science, we recall, is our attempt to understand the reproducible and predictable aspects of nature, and thus of perceived reality. (Indeed, in the section *Perceived Reality* of this chapter I pointed out that science is concerned with perceived reality.) Both nature and perceived reality are phenomena of objective reality. Objective reality transcends and subsumes nature.

Earlier in this book, in the section *Transcendence and Nontranscendence* of chapter 3, I declared that I hold a nontranscendent world view, which is a world view that does without any reality transcending nature. In the present chapter we saw how science guides us to a world view of partially hidden objective reality, and as might be guessed, I subscribe to that world view. Since objective reality transcends nature, have I then run into a contradiction? Yes, indeed I have. At the stage of our investigation that was discussed in that section of chapter 3 I wanted to hold as simple a world view as seemed to be warranted by science, thus a non-transcendent world view. I assumed, as do many scientists, that we would thereby find the objectivity we strive for through science. Transcendence smacked too much of subjectivity.

But it turned out, as we saw in the present chapter, that science, in its study of nature, cannot fulfill our demand for objectivity. The quantum aspect of nature involves observers too much for that. So any objective reality must be "farther" from us than nature, than perceived reality. It must transcend them. Thus I am led to the belief in partially hidden objective reality, a reality transcending nature, a reality most likely surpassing human understanding. But an objective reality! An observer-independent underlying reality that is the same reality for protons, for planets, for peanuts, and for people.

Is there room there for God? Deities? Omniscience? Omnipotence? Supernatural powers? Grand designs? There is indeed. (Instead, I was tempted to write: I am afraid so.) Once Pandora's box of transcendence is opened, almost anything goes. After all, we are talking about a metaphysical position (see the section *Science and Metaphysics* of chapter 3), albeit a "scientific" one, so one is free to assume whatever one wants, as long as consistency with science is maintained. And, let me emphasize, as long as objectivity is maintained: as long as one's assumptions do not contradict observer-independence, or the idea that it is the same underlying reality for helium, for helicopters, for hyacinths, and for humans. If objectivity were to be abandoned, there would be no point in the whole metaphysical construct we developed in this chapter. And self-consistency? As one likes, since we do not really expect fully to understand objective reality anyhow.

Reality

SUMMARY

Using arguments and justifications based on common sense, on our understanding of nature, and on reasonableness, utility, conceptual economy, and motivation, we are led by science to the belief in an objective reality. Our scientific knowledge of that reality, however, is indirect. But science should nevertheless give us a literal description of objective reality, at least for those aspects of it that strongly and frequently affect our survival as individuals and as a species. Yet quantum theory cannot be a literal description of that reality. Thus objective reality is partially, very likely mostly, hidden from us. We should not exclude the possibility that other, nonscientific modes of comprehension and understanding might offer insight into objective reality. Since that reality transcends nature, we are thus led to a transcendent world view.

BIBLIOGRAPHY

The principal references for this chapter are B. d'Espagnat, *In Search of Reality* (New York: Springer-Verlag, 1983), which is not too easy reading; and B. d'Espagnat, *Reality and the Physicist: Knowledge, Duration and the Quantum World* (Cambridge: Cambridge University Press, 1989), which is still harder, even for theoretical physicists. This chapter is in essence an adaptation, with twists of my own, of d'Espagnat's ideas concerning what he aptly calls veiled objective reality. That will become obvious upon reading his book(s).

For more books dealing in various ways with reality, see: E. Harrison, *Masks of the Universe* (New York: Macmillan, 1985); R. Morris, *Dismantling the Universe: The Nature of Scientific Discovery* (New York: Simon and Schuster, 1983); R. Morris, *The Nature of Reality* (New York: McGraw-Hill, 1987); as well as chapter 1 of R. K. Adair, *The Great Design: Particles, Fields, and Creation* (Oxford: Oxford University Press, 1987); chapter 16 of M.

Eigen and R. Winkler, *Laws of the Game: How the Principles of Nature Govern Chance* (New York: Knopf, 1981); chapter 16 of H. Fritzsch, *The Creation of Matter: The Universe from Beginning to End* (New York: Basic Books, 1984); the "epistemological prolegomena" of K. Lorenz, *Behind the Mirror: A Search for a Natural History of Human Knowledge* (New York: Harcourt Brace Jovanovich, 1977); chapter 3 of D. Park, *The How and the Why* (Princeton, N.J.: Princeton University Press, 1988); chapter 10 of A. Rae, *Quantum Physics: Illusion or Reality* (Cambridge: Cambridge University Press, 1986); chapter 11 of J. S. Trefil, *Reading the Mind of God: In Search of the Principle of Universality* (New York: Charles Scribner's Sons, 1989); and chapter 12 of A. Zee, *Fearful Symmetry: The Search for Beauty in Modern Physics* (New York: Macmillan, 1986).

For adaptive biological evolution, see: Lorenz, above; chapter 5 of Eigen and Winkler, above; and chapter 7 of J. Monod, *Chance and Necessity: An Essay on the Natural Philosophy of Modern Biology* (New York: Knopf, 1971).

For the observer dependence of quantum reality, see: chapter 10 of Adair, above; chapter 3 of J. D. Barrow, *The World Within the World* (Oxford: Oxford University Press, 1988); chapter 8 of P. C. W. Davies, *God and the New Physics* (New York: Simon and Schuster, 1983); chapter 4 of P. C. W. Davies, *Other Worlds: A Portrait of Nature in Rebellion: Space, Superspace and the Quantum Universe* (New York: Simon and Schuster, 1980); chapter 1 of P. C. W. Davies and J. R. Brown, eds., *The Ghost in the Atom: A Discussion of the Mysteries of Quantum Physics* (Cambridge: Cambridge University Press, 1986); chapters 10–13 of part 1 of H. R. Pagels, *The Cosmic Code: Quantum Physics as the Language of Nature* (New York: Simon and Schuster, 1982); chapters 5–8 of J. C. Polkinghorne, *The Quantum World* (Princeton, N.J.: Princeton University Press, 1984); chapters 3–5, 10 of Rae, above; and chapter 11 of F. Rohrlich, *From Paradox to Reality: Our New Concepts of the Physical World* (Cambridge: Cambridge University Press, 1987).

9 Self-generating Universe

...
Universe born of itself,
Looping through being-becoming.
A Planckian itch, a space-time twitch,
A bitch of a glitch, a world is born.
THE WORLD IS BORN!
Universe born of itself,
Looping through being-becoming.
...

COSMOLOGICAL SCHEMES

In the section *Capricious Cosmos* of chapter 4 I briefly presented two cosmological schemes that have been proposed as bringing about the existence of ensembles of universes, where we saw that this "existence" is merely in a metaphysical sense, outside the framework of science. Let us now consider those schemes in themselves. Schemes of the big bang type have the universe come into being in a cosmic explosion, a primeval fireball of extreme density, pressure, and temperature, following which it expands and cools. The eventual fate of the universe is either eternal expansion or expansion only to some maximal extension, after which the universe contracts down again and ends its life in a cosmic implosion, the big crunch.

If it is assumed that the universe eventually collapses, then a "cosmic oscillation" scheme can be constructed (metaphysically, of course) by concatenating such collapsing universes, i.e., by picturing a scenario in which universe follows universe forever: Bang, expansion, contraction, crunch, bang, expansion, contraction, crunch.... The concepts "follows" and "forever" here must be assigned some metaphysical sense, since they cannot have their usual meanings. Time is connected with the evolution of the universe; it is an aspect of the universe. Time has meaning only "inside" the universe. But the universes of the scheme do not follow each other "inside" the universe. The relation of one universe to another in this scheme is "outside" time; it is not temporal. The scheme must be pictured as taking place in some metatime, some "higher" time, unrelated to universe time. In metatime we then have the scenario of universe following universe, where each universe comes into being via a big bang, expands and contracts in its own time, and perishes in a big crunch.

Now, it has been assumed that only eventually collapsing universes can take part in the cosmic oscillation scheme. Eternally expanding ones have been banned because of the reasoning that, were any universe in the sequence to expand forever, there would be no next universe—its turn would never arrive. That discrimination is a result of conceptually mixing up universe time, the private, internal time of each universe, and metatime, the metaphysical time in which the universes are considered to follow each other. Those two kinds of time are absolutely and totally unrelated. (And so, for that matter, are the universe times of different universes in the scheme.) So there is really no reason that eternally expanding universes cannot join the party too. So what if they live forever? Forever is in their own internal times. In metatime they can be assigned whatever metalifetimes one pleases. Then clear the stage for the next act.

The big bang cosmological scheme sorely tempts us with questions of what was before the big bang and what will be after the big crunch, if indeed the universe will collapse. Recognizing that time is connected with the evolution of the universe, we realize that such questions amount to asking what happened before time's beginning and what will happen after time's end. But the beginning and end of time have no meaning in relation to time itself. To attach any significance to those concepts one must put oneself "outside" of time, i.e., go to a metatime in which natural time is "imbedded." Thus questions of before the big bang and after the big crunch are not questions about nature, but rather are metaphysical questions. Still, if we have not yet thoroughly internalized the difference between science and metaphysics, we are tempted to ask such questions as if science had something meaningful to say on the subject. It has not.

Such questions make no more science sense for the oscillatory scheme than they do for the single shot scheme. But in the former case they can be endowed with metaphysical sense, if only they are understood as referring to metatime, not to universe time. Before (referring to metatime) the big bang was (in metatime) the big crunch or eternal expansion of the previous universe, and after (in metatime) the big crunch or eternal expansion will come (in metatime) the big bang of the next universe.

The scheme can be made compact and conceptually more economical (i.e., requiring less conceptual raw material) by metaphysically identifying the final state of the universe with its own initial state. That would certainly be easier to do conceptually in the case of the universe collapsing to a big crunch than in the case of it eternally expanding, because a big crunch seems ripe for bursting into a big bang. (But such considerations are pure metaphysics, as picturesque as they might be.) Then "before the big bang" would simply mean before the big crunch, and "after the big crunch" would be after the big bang. That sounds like "closed" time, where the end brings us back to the beginning. But again, the beginning and end of time have no meaning in relation to time itself; they are beyond nature. Thus the "closed" character of time in that scheme is only in relation to metatime. There the scheme can indeed be viewed as closed time. Alternatively it can be pictured as the universe repeating itself forever, which might be somewhat easier to picture. The conceptual economy of the closed time scheme is that only a single universe, possibly along with an infinity of its replicas, is involved, rather than an infinity of different universes.

We might also be tempted to ask Why? Why did the universe come into being? Why is the universe as it is and not different? And if we have not thoroughly internalized the lesson of this book, especially of the section *Capricious Cosmos* of chapter 4, we might even be tempted to expect science to answer our questions. But, as we have seen, such questions cannot be answered meaningfully within the framework of science. An answer to such a question would be a law of behavior for the universe, and we have seen that the universe is inherently lawless.

BABY UNIVERSES

I would like to present a new big bang–type scheme. That scheme involves an ensemble of universes, yet it does not impose laws of behavior for universes so it reduces the temptation to ask Why? It even answers a question of "before." Moreover, the proposed scheme is very conservative (makes a minimum of new assumptions) and conceptually economical (uses a minimum of conceptual raw material). Before proceeding, let me emphasize that the scheme is *metaphysical*, just like other cosmological schemes. It is a way of looking at things. *It is not a theory.*

In the standard big bang–type schemes, as described above, the coming into being of the universe and its possible going out of existence are both beyond nature. The first step in the construction of our scheme is to make the conservative assumption that, without (yet) making it part of nature, the coming into existence of the universe is governed by the same laws of nature that we find within the universe here and now, specifically, the quantum laws of nature. (That certainly sounds like a law of behavior for the universe, but let us put it on hold for the meantime and await further developments later on.)

In the section *Capricious Cosmos* of chapter 4, in the sections *Observer and Observed* and *Quasi-isolated System and Surroundings* of chapter 5, and in the section *Whence Order?* of chapter 6 I mentioned some of the quantum characteristics of nature: fundamental nonseparability, uncontrollable correlations, inherent unpredictability, and probabilistic (rather than deterministic) behavior, all of which are significant mainly at the submicroscopic level. In addition to all that, quantum theory also tells us that there is no quiescence in nature, that even the most nearly perfect vacuum is alive and bustling with submicroscopic activity. Subnuclear particles are constantly and spontaneously materializing out of nothing and dematerializing to nothing, while during their fleeting lifetimes they interact with each other in many and various ways.

Now, although this point is less well understood, not even space and time should be immune to such unrest. All over space the very space-time fabric of the universe should be in constant submicroscopic turmoil, picturesquely described as seething space-time foam. The exact description of that is very complicated and technical. Some authors attempt to impart a feeling for it with the help of diagrams containing curved and wavy lines, indicating the distortion of space-time. I do not see much point in that and will try to do the job verbally. The idea is that at a sufficiently tiny scale, an unimaginably tiny scale called the Planck scale (after Max Planck, 1858–1947) and characterized by lengths of the order of 10^{-35} meters and time intervals of the order of 10^{-43} seconds, space and time should be found to lose their ordinary meaning. At that scale we should find that in a random and highly fluctuating way lengths and time intervals become ill-defined, direction loses its significance, past and future blend, and even the distinction between space and time becomes blurred. The situation is termed quantum fluctuations of space-time.

Within that agitation there is supposed to be incessant and spontaneous closing up of Planck-size regions of space-time upon themselves, whereby they become disconnected from the universe, forming distinct and separate universes in themselves, called baby universes. Thus the universe, thanks to its quantum character, is thought to give birth to baby universes. The contents of baby universes are totally and absolutely inaccessible to us, since they are disconnected from the universe. Once formed, baby universes cannot be thought of as having any location in the universe nor of participating in our time. Their mode of existence is external to the material world, thus beyond nature and outside the domain of concern of science. Baby universes are metaphysical entities.

Now the first step in the construction of our cosmological scheme, to be more specific than I was above, is to assume that the same quantum laws that govern baby universe production within the universe govern the coming into being of the universe itself. In other words, the universe is assumed to be a baby universe.

CLOSING THE CIRCLE

So the baby universes, which our understanding of quantum theory as applied to space-time tells us are born of the universe, are thus assumed to be born of what is itself a baby universe, which is then supposed to be born of . . . what? We seem to be led to an infinite regress of baby universes born of baby universes born of baby universes, and so on and on. But we put an end to this infinity by the second step in the construction of our scheme: We conceptually identify the baby universe that is the universe with a baby universe born of *the universe itself*. Thus we impose circularity and self-reference rather than infinity. That is the conceptual economy of the scheme. The universe gives birth to itself, generates itself. That is the self-generating universe scheme.

The self-generating universe scheme involves no special law of behavior for the universe. The coming into being of the universe

Self-generating Universe

is governed by the very same laws of nature that are valid within the universe itself in a manner that seems very natural. The universe controls its own birth through the very laws of nature that are valid for aspects and phenomena of itself. Its birth is a phenomenon of the universe itself. The universe generates itself.

Note that the self-generating universe scheme is just that, a scheme, a picture. *It is not a theory.* It is a way of viewing things. It is a conceptual framework upon which we can hang our ideas and that can guide us in our theorizing. It is a conceptual vessel to contain, constrain, and combine our theories. It cannot be confirmed or falsified in the manner of scientific confirmation and falsification of theories. It can, however, be judged for its consistency with science and its self-consistency and for its effectiveness, enlightening, elegance, and economy.

Note also that the identification of the universe with a baby universe born of itself is a *conceptual* identification only. It has no material significance, as the mode of existence of baby universes is external to the material world. Since baby universes are disconnected from the universe, that identification does not imply that the universe contains itself in any way. The self-*generating* universe scheme is not a scheme of a self-*containing* universe.

The fact that the sizes and time intervals that characterize the universe are not all unimaginably tiny, on the order of 10^{-35} meters and 10^{-43} seconds, as might be expected from the assumed Planck-scale origin of the universe as a baby universe born of itself, is explained by the absolute inaccessibility of the contents of baby universes from their mother substrate and the absolute inaccessibility of the mother substrate from within baby universes. No measurement standard, no standard meter stick or clock ticking off standard seconds, can in principle be passed between them. Thus sizes and durations as perceived in the universe, as a baby universe, bear no relation at all to any sizes and durations that might be thought of in any sense as existing "outside it," even when "outside it" is within itself according to the self-generating universe scheme. So there is no contradiction between the 10^{-43}

second Planck duration and the fact that the universe seems to have existed for considerably more than a thousand million years nor between the 10^{-35} meter Planck size and the gigantically larger size of the universe.

Of all the baby universes born of the universe, which one is the universe itself? One being formed and disconnecting from the universe right now in my left ear? Or one that was born who knows where millions of years ago? Or one that will be born eons and eons from now? We cannot know. First of all, the contents of a baby universe are inaccessible to us. Second of all, even if we *imagined* we could peer into baby universes and see what is going on there, we still would not be able to pick out the one that is our reality. That follows from the fundamental unpredictability and indeterminacy of nature as described by quantum theory.

More precisely, just as our reality consists of our past and our present and its emanating quantum *possibilities* for the future, so we assume in the self-generating universe scheme that we would find an unimaginable number of baby universes containing the same past and present as ours and with all possible futures. (We would also find an unimaginable number of baby universes containing conceivably alternative pasts and presents to ours, as well as those containing nothing resembling our reality at all.) Thus over and above the inaccessibility to us of the contents of baby universes, we have this uncertainty in picking out the one that is the universe, the uncertainty that is just the quantum uncertainty of our future.

MANY WORLDS

So one of the baby universes born of the universe is identified with the universe itself. What about the unimaginable number of baby universes that are not the universe? For self-consistency of the self-generating universe scheme those other baby universes must also be identified with universes, and science has a perfect role for such a multitude of universes: the plethora of quantum possibilities. We

Self-generating Universe

thus obtain a framework for the many worlds interpretation of quantum theory, as mentioned in the section *Capricious Cosmos* of chapter 4, whereby at every instant the universe "branches" into realizations of all the quantum possibilities of that instant, which continue to "coexist side-by-side," each branch universe branching further at the next instant, and so on to unimaginable multitude.

The self-generating universe scheme allows us, in a sense, to picture the existence of those coexisting, side-by-side universes and to relate them to our reality: They are all baby universes born of the universe itself. Here lie all the "discarded" universes, all those failed quantum possibilities, all possible realities that are not ours. They are not our reality, but they closely accompany our reality in an essential way. They are generated by our reality just as our reality is generated by itself, by means of quantum space-time fluctuations producing baby universes.

That is the self-generating universe scheme. It possesses a beautiful completeness: The universe is the cause and effect of its own existence (as well as of all its quantum alternatives), a self-generating unity. Thus the picture is very economical, as it exploits a minimum of conceptual raw material. Within the completeness of the self-generating universe scheme conventional and anthropic explanations can be carried out just as in the simpler picture assumed in previous chapters. An explanation of the whole business, as before, cannot be provided by science.

The picture provides an answer to the age-old question of what "was before" the coming into being of the universe (also called "the Creation"). The universe was preceded by and evolved from a quantum fluctuation of space-time, whereby a Planck-size region of space-time closed up upon itself, became disconnected from the universe, and formed a distinct and separate universe in itself, the baby universe that is the universe.

Now here is an interesting point concerning time. Ordinarily the question "What was before the coming into being of the universe?" is well considered meaningless within science, as long as "was before" is meant in the literal temporal sense of "preceded in time." That is because time is an aspect of the universe rather than the universe existing in time, as was mentioned earlier in the present chapter. Nevertheless, the self-generating universe scheme attributes meaning to that question by having time serve also as metatime, in accord with the self-referential character of the scheme. It seems meaningful to claim that a quantum space-time fluctuation occurs around some time and that such a fluctuation precedes the formation of a baby universe. But the occurring and preceding are with respect to the time of the universe and have nothing to do with any internal time the baby universe might possess. When the universe is identified with a baby universe born of itself, it can then be said to have been preceded by a space-time fluctuation and its birth to have occurred immediately following the occurrence of that fluctuation. The preceding, following, and occurring are with respect to the time of the universe considered as a metatime in which (baby) universes, including the universe itself, form.

That reasoning must seem quite tortuous. In fact it is, due to the circularity inherent to any self-referential situation. And the self-generating universe scheme is self-referential. It takes getting used to.

SUMMARY

A very conservative and conceptually economical big bang–type cosmological scheme, the self-generating universe scheme, is proposed, according to which the universe as well as all its quantum alternatives are identified with baby universes born of the universe itself. Thus the coming into existence of the universe is governed by the same laws of nature that are valid within the universe, and a framework is obtained for the many worlds interpretation of quantum theory. The universe is viewed as the cause and effect of its own existence (as well as of all its quantum alternatives).

BIBLIOGRAPHY

For cosmological schemes implying the existence of ensembles of universes, see: chapters 2 and 5 of J. D. Barrow and J. Silk, *The Left Hand of Creation: The Origin and Evolution of the Expanding Universe* (London: Heinemann, 1983); chapter 11 of P. C. W. Davies, *The Runaway Universe* (New York: Harper and Row, 1978); part 3 of J. Gribbin and M. Rees, *The Stuff of the Universe: Dark Matter, Mankind and the Coincidences of Cosmology* (London: Heinemann, 1989); chapter 12 of R. Morris, *The End of the World* (Garden City, N.Y.: Anchor Press, 1980); and chapter 8 of R. Morris, *The Fate of the Universe* (New York: Playboy Press, 1982).

For the quantum vacuum, space-time foam, quantum fluctuations, universe production, and such matters, see: chapters 12 and 18 of R. K. Adair, *The Great Design: Particles, Fields, and Creation* (Oxford: Oxford University Press, 1987); chapter 4 of J. D. Barrow, *The World Within the World* (Oxford: Oxford University Press, 1988); chapters 2 and 3 of Barrow and Silk, above; chapter 8 of P. C. W. Davies, *The Edge of Infinity* (New York: Simon and Schuster, 1981); chapter 16 of P. C. W. Davies, *God and the New Physics* (New York: Simon and Schuster, 1983); chapters 3 and 9 of P. C. W. Davies, *Other Worlds: A Portrait of Nature in Rebellion: Space, Superspace and the Quantum Universe* (New York: Simon and Schuster, 1980); chapter 8 of S. W. Hawking, *A Brief History of Time: From the Big Bang to Black Holes* (London: Bantam, 1988); chapter 9 of T. Hey and P. Walters, *The Quantum Universe* (Cambridge: Cambridge University Press, 1987); chapter 9 of Morris, *The End of the World*, above; chapter 8 of part 2 of H. R. Pagels, *The Cosmic Code: Quantum Physics as the Language of Nature* (New York: Simon and Schuster, 1982); chapters 4 and 5 of part 3 of H. R. Pagels, *Perfect Symmetry: The Search for the Beginning of Time* (New York: Simon and Schuster, 1985, and Toronto: Bantam, 1986); chapter 6 of J. Silk, *The Big Bang*, revised and updated ed. (San Francisco: Freeman, 1989); chapter 13 of J.

S. Trefil, *The Moment of Creation: Big Bang Physics from Before the First Millisecond to the Present Universe* (New York: Charles Scribner's Sons, 1983); and chapter 15 of J. S. Trefil, *Reading the Mind of God: In Search of the Principle of Universality* (New York: Charles Scribner's Sons, 1989).

For the many worlds interpretation of quantum theory, see: P. C. W. Davies and J. R. Brown, eds., *The Ghost in the Atom: A Discussion of the Mysteries of Quantum Physics* (Cambridge: Cambridge University Press, 1986); chapter 3 of Barrow, above; chapter 5 of P. C. W. Davies, *The Accidental Universe* (Cambridge: Cambridge University Press, 1982); chapter 13 of part 1 of Pagels, *The Cosmic Code*, above; and chapter 6 of A. Rae, *Quantum Physics: Illusion or Reality* (Cambridge: Cambridge University Press, 1986).

Glossary

Terms in CAPITAL LETTERS can be found in the Glossary.

anthropic principle
The existence of *Homo sapiens* may, within the framework of SCIENCE, serve as an explanation for phenomena and aspects of NATURE, and moreover, such explanations are the most fundamental.

baby universe
According to the ideas of QUANTUM THEORY, the very SPACE-TIME fabric of the universe should be in submicroscopic turmoil, a seething space-time foam. Within that agitation there is incessant and spontaneous closing up of unimaginably tiny, PLANCK-SCALE regions of space-time upon themselves, whereby they become disconnected from the universe, forming distinct and separate universes in themselves, called baby universes. The contents of those universes are inaccessible to us. Baby universes are metaphysical [METAPHYSICS] entities.

beauty
As referred to a THEORY, a beautiful theory is one that arouses the esthetic feeling of beauty in the (initiated) scientist considering it. The feeling of beauty is found to be engendered by

the properties of simplicity, unification, and generality, when possessed by that theory. Scientists, in their belief that NATURE should be understandable in terms of beautiful theories, tend to prefer more beautiful theories over less beautiful ones, even at the expense of objective advantages. Amazingly, successful theories do tend to be beautiful.

big bang
The cosmic explosion, or primeval fireball, by which, according to certain cosmological SCHEMES, the universe came into existence and has been expanding ever since.

big crunch
The cosmic implosion, or utter collapse, by which, according to certain cosmological SCHEMES, the universe will go out of existence after passing through a contraction era following the present expansion era.

conservatism
In SCIENCE: Hold on to what you have, stick to the tried and well confirmed, for as long as is reasonably possible; make changes only when the need for change becomes overwhelming; and then make only the minimal changes needed to achieve the desired end. Specifically with regard to the LAWS OF NATURE: As long as there is no compelling reason to the contrary, assume that the laws of nature we find here and now are, were, and will be valid everywhere and forever.

cosmological scheme
See SCHEME.

cosmology
The study of the working of the cosmos, the universe as a whole, at present, in the past, and in the future. In its dealings with the connections and interrelations among the aspects and

phenomena of the universe, cosmology can be considered a branch of SCIENCE. But in its holistic [HOLISM] mode, when it attempts to comprehend the universe as a whole, it is a branch of METAPHYSICS.

dimension
For our discussion, the possibility of assigning a measure to something.

elementary particle
Any of many kinds of subatomic and subnuclear particles, including the electron, positron, proton, neutron, neutrino, pion, etc. The name is historic, as the actual elementarity of many of the so-called elementary particles is dubious, according to present-day understanding, which views them as being composed of more elementary constituents.

evolution (of nature)
The process of NATURE's change in TIME.

evolution, law of
See LAW OF EVOLUTION.

extended Mach principle
The origin of the LAWS OF NATURE for QUASI-ISOLATED SYSTEMS lies with the universe as a whole.

falsifiability
Testability. The property of a THEORY that it can be tested against as yet unknown natural [NATURE] phenomena to determine whether it is true or false. In order to be falsifiable a theory must predict something in addition to what it was originally intended to explain.

holism
The metaphysical [METAPHYSICS] position that NATURE can be understood only in its wholeness, including human beings, or not at all. (Compare with REDUCTIONISM.)

idealism

The metaphysical [METAPHYSICS] position that the LAWS OF NATURE are not inherent to the external world, to OBJECTIVE REALITY, but are mental constructs, artifacts of the way our minds interpret and organize our sensory impressions, of the way we perceive the world. (Compare with REALISM.)

inertia

The property of bodies according to which a body's behavior is governed by Newton's first universal law of motion [NEWTON'S LAWS]: In the absence of forces acting on them or when such forces cancel each other, bodies remain at rest or continue to move uniformly in a straight line. (See MACH PRINCIPLE.)

inflation

According to certain cosmological SCHEMES, an era of unimaginably rapid expansion of the universe (called the inflationary era) starting soon after the BIG BANG, during which era the expanding universe broke up into island universes, one of which is ours.

initial state

Any state of a material system, when considered as a precursor state from which the subsequent EVOLUTION of the system follows. (See LAW OF EVOLUTION.)

isolated system

The idealization of a material system that has absolutely no interaction with the rest of the universe. An isolated system would not be part of NATURE, since we could not interact with it and hence could not observe it. (Compare with QUASI-ISOLATED SYSTEM.)

Kepler's laws

Johannes Kepler's three laws [LAW] of planetary motion:

Glossary

(1) The path each planet traverses in SPACE, its orbit, lies wholly in a fixed plane and has the form of an ellipse, of which the Sun is located at a focus.

(2) As each planet moves along its elliptical orbit, the (imaginary) line connecting it with the Sun sweeps out equal areas during equal time intervals.

(3) The ratio of the squares of the orbital periods of any two planets equals the ratio of the cubes of their respective orbital major axes.

These laws express an ORDER among the astronomical data on the planets and offer a description and a unification of them. They predicted the relevant properties of the motions of the planets discovered after Kepler's time. They are also valid for any system of astronomical bodies revolving around a massive central body, such as the moons of Jupiter.

> **law (of nature)**
> An expression of ORDER, thus of simplicity, in NATURE. A compact condensation of all relevant existing data, as well as of any amount of potential data, a law is a unifying, descriptive device for its relevant class of natural phenomena. Laws enable us to predict the results of new experiments. (See PREDICTABILITY.)

> **law of evolution**
> Any LAW OF NATURE that, given any INITIAL STATE of a QUASI-ISOLATED SYSTEM, gives the state that evolves [EVOLUTION] from it at any subsequent TIME.

> **law of nature**
> See LAW.

> **laws, Kepler's**
> See KEPLER'S LAWS.

147

laws, Newton's
See NEWTON'S LAWS.

Mach principle
The origin of INERTIA lies with all the matter of the universe.

Mach principle, extended
See EXTENDED MACH PRINCIPLE.

many worlds
The many worlds interpretation of QUANTUM THEORY holds that at every instant the universe "branches" into realizations of all the quantum possibilities of that instant, which continue to "coexist side-by-side," each branch universe branching further at the next instant, and so on.

metaphysics
A branch of philosophy dealing with being and reality. Metaphysics, as used in this book, is the philosophic framework in which SCIENCE operates. In that sense metaphysics is concerned with what lies around, below, above, before, and beyond science. A metaphysical position is part of one's WORLD VIEW.

metatime
A metaphysical [METAPHYSICS] time "higher" than ordinary time, in which universes can be born and die.

nature
The material universe with which we can, or can conceivably, interact; i.e., everything of purely material character that we can, or can conceivably, observe and measure. By "conceivably" we mean that it is not precluded by any principle known to us and is considered attainable through further technological research and development. We exclude from nature such concepts as mind, idea,

feeling, consciousness, etc., since the question of their character —material or otherwise—has not been settled.

Newton's laws
Isaac Newton's three universal laws [LAW] of motion:

(1) In the absence of forces acting on it or when such forces cancel each other, a body will remain at rest or continue to move uniformly in a straight line.

(2) A force acting on a body will cause the body to undergo acceleration whose direction is that of the force and whose magnitude is proportional to that of the force divided by the body's mass.

(3) For every force acting on it a body will react upon the force's source with a force of opposite direction and equal magnitude along the same line of action.

Newton's law of universal gravitation: Every pair of bodies undergoes mutual attraction, with the force acting on each body proportional to the product of the bodies' masses and inversely proportional to the square of their separation.

These four laws form Newton's THEORY to explain KEPLER'S LAWS of planetary motion (as well as a vast realm of other mechanical phenomena).

nontranscendent world view
See WORLD VIEW, NONTRANSCENDENT.

objective reality
OBSERVER-independent reality underlying the phenomena of NATURE. SCIENCE shows that objective reality must be partially, very likely mostly, hidden from us. (See PERCEIVED REALITY.)

observed
The rest of NATURE, as distinct from us (*Homo sapiens*) as OBSERVER, in the observer-observed separation [SEPARABILITY]

of nature, according to the reductionist [REDUCTIONISM] approach.

observer
We (*Homo sapiens*), as distinct from the rest of NATURE as OBSERVED, in the observer-observed separation [SEPARABILITY] of nature, according to the reductionist [REDUCTIONISM] approach.

order
The opposite of randomness, of haphazardness. The existence of relations among natural [NATURE] phenomena. Order in nature is a simplicity of nature.

particle, elementary
See ELEMENTARY PARTICLE.

perceived reality
That which we become aware of via our physical senses. Perceived reality is the actual subject of SCIENCE. (See OBJECTIVE REALITY.)

phenomenon, unique
See UNIQUE PHENOMENON.

Planck scale
The scale of lengths and time intervals (durations) at which SPACE and TIME exhibit quantum [QUANTUM THEORY] characteristics: about 10^{-35} meters and 10^{-43} seconds.

positivism
The metaphysical [METAPHYSICS] position that only our sense data, derived from measurements and observations, are fundamental. (Compare with REALISM.)

predictability

The characteristic that among the natural [NATURE] phenomena investigated, ORDER can be found, from which LAWS can be formulated, predicting the results of new experiments. Predictability makes SCIENCE a means both to understand and to exploit nature. We do not claim that nature is predictable in all its aspects, but any unpredictable aspects it might possess lie outside the domain of science.

quantum theory

A very formal and mathematical THEORY concerned with the fundamental behavior of all material systems in principle, but usually and most usefully applied to molecular, atomic, and subatomic systems. Quantum theory is formulated in terms of possible happenings and their probabilities of actually occurring (rather than in strictly deterministic terms of what *will* occur). Individual submicroscopic events, according to this theory, are inherently unpredictable [PREDICTABILITY]; it is only their probability that can be predicted. Quantum theory implies, among other things, that NATURE is fundamentally nonseparable [SEPARABILITY], so that uncontrollable correlations can exist among separated systems. Quantum theory also implies that PERCEIVED REALITY IS OBSERVER-dependent, thus that perceived reality cannot be a literal description of OBJECTIVE REALITY. Another quantum effect is the constant submicroscopic activity occurring even in the most nearly perfect vacuum, in which turmoil even SPACE and TIME themselves should take part. (See MANY WORLDS, BABY UNIVERSE.)

quasi-isolated system

Any material system that is as nearly isolated as possible from the rest of the universe, i.e., whose interaction with the rest of the universe is reduced to the minimum possible. (Compare with ISOLATED SYSTEM; see SURROUNDINGS.) It is for quasi-isolated systems that LAWS OF NATURE are found.

realism

In general, the metaphysical [METAPHYSICS] position that there exists an underlying OBSERVER-independent, OBJECTIVE REALITY, that NATURE would manage just as well if we were not around. Also specifically, the metaphysical position that the LAWS OF NATURE reside in that reality. (Compare with POSITIVISM and with IDEALISM.)

reality, objective
See OBJECTIVE REALITY.

reality, perceived
See PERCEIVED REALITY.

reductionism

The metaphysical [METAPHYSICS] position that NATURE can be understood as the sum of its parts and thus should be studied by analysis and synthesis. (Compare with HOLISM; see SEPARABILITY.)

reproducibility

The possibility of repeating experiments by the same and other investigators, thus giving data of objective, lasting value about the phenomena of NATURE. Reproducibility makes SCIENCE a common human endeavor, and as nearly as possible, an objective endeavor of lasting validity. Nature is not claimed to be reproducible in all its aspects, but any irreproducible aspects it might possess lie outside the domain of science.

scheme

Especially cosmological [COSMOLOGY] scheme, an attempt to *describe* the working of the cosmos, the universe as a whole. Cosmological schemes are not THEORIES, do not *explain* the working of the cosmos, since the universe as a whole, being a UNIQUE

PHENOMENON and thus irreproducible [REPRODUCIBILITY], lies outside the domain of SCIENCE.

science
Our attempt to understand, i.e., to be able to explain, the reproducible [REPRODUCIBILITY] and predictable [PREDICTABILITY] aspects of NATURE. Science is a human endeavor, since *Homo sapiens* plays an essential role in it in that: (1) science is a by-product of *our* existence; (2) it is *our* conception of nature that we are attempting to explain; and (3) a valid explanation in science is one that satisfies *us*.

self-generating universe
A very conservative [CONSERVATISM] and conceptually economical BIG BANG–type cosmological SCHEME, according to which the universe as well as all its quantum [QUANTUM THEORY] alternatives are identified with BABY UNIVERSES born of the universe itself. Thus the coming into being of the universe is governed by the same LAWS OF NATURE that are valid within the universe, and a framework is obtained for the MANY WORLDS interpretation of quantum theory. The universe is viewed as the cause and effect of its own existence (as well as of all its quantum alternatives).

separability
The amenability of NATURE to our attaining understanding of it through analysis and synthesis. REDUCTIONISM holds that separability is valid for nature.

space
The DIMENSION of being.

space-time
SPACE and TIME together, considered as a single concept. Einstein's THEORIES of relativity are formulated in terms of space-time.

surroundings (of quasi-isolated system)
The rest of the universe from any QUASI-ISOLATED SYSTEM.

system, isolated
See ISOLATED SYSTEM.

system, quasi-isolated
See QUASI-ISOLATED SYSTEM.

theory
A scientific [SCIENCE] explanation of a LAW OF NATURE. A theory gives reasons for the law it explains. The acceptability of a theory depends on its giving the feeling that something is indeed being explained. That feeling is found to be fostered by the existence of a number of properties of a theory, the essential one being that whatever is explaining must logically imply that which is being explained. Additional properties enhancing a theory's acceptability are that what is explaining be as much an aspect of nature as what is being explained and that the former be more general, more fundamental, more unifying, and simpler than the latter and be perceived as causing the latter. Beautiful [BEAUTY] theories are preferred. A theory should be falsifiable [FALSIFIABILITY].

theory, quantum
See QUANTUM THEORY.

Theory of Everything (TOE)
The general name for any putative THEORY explaining the universe as a whole. However, a TOE, were one to be devised, could not be a theory, could not be an *explanation*, since the universe as a whole lies outside the domain of SCIENCE. It would be a SCHEME, a *description* of the universe as a whole.

time
The DIMENSION of becoming.

transcendent world view
See WORLD VIEW, TRANSCENDENT.

unique phenomenon
A natural [NATURE] phenomenon whose special character makes it *essentially* different from all other natural phenomena. Thus a unique phenomenon is inherently irreproducible [REPRODUCIBILITY]. The universe as a whole is an example *par excellence* of a unique phenomenon.

world view
One's attitude toward and interpretation of reality; the conceptual framework by which one organizes one's perceptions.

world view, nontranscendent
Any WORLD VIEW that does without a reality beyond, or transcending, NATURE and makes do with nature as all there is. Nontranscendent world views do not necessarily deny the existence of mind, consciousness, thought, emotion, feeling, etc. (Compare with WORLD VIEW, TRANSCENDENT.)

world view, transcendent
Any WORLD VIEW involving the existence of a reality beyond, or transcending, NATURE. Nature, with which SCIENCE is concerned, is viewed as being embedded in, being part of, that transcendent reality. A religion is an example of a transcendent world view. (Compare with WORLD VIEW, NONTRANSCENDENT.)

Index

abstraction, 13
acceptability of theory, *see* theory, acceptability of
acceptable theory, *see* theory, acceptable
acceptance of theory, *see* theory, acceptance of
adaptive evolution, *see* evolution, adaptive
age of the universe, *see* universe, age of the
alternative, quantum, *see* quantum alternative
analogy, 3, 39, 90

analysis, 56, 67, 69, 70 77–81, 88, 152–153
anthropic explanation, *see* explanation, anthropic
anthropic principle, *see* principle, anthropic
anti-isolation, 75–77, 82–83
archetypical example, *see* example, archetypical
aspect, quantum, *see* quantum aspect
aspect of nature, *see* nature, aspect of

Index

aspect of reality, *see* reality, aspect of
aspect of the universe, *see* universe, aspect of the
astrology, 65
astronomer, 56
astronomical scale, *see* scale, astronomical
astronomy, 14–15, 30, 53–54, 70, 85, 87, 92–93, 102, 147
atom, 8, 27–28, 49, 68, 70–72, 91–93, 104, 119, 151
attraction, 24–25, 30–31, 149
awareness, 69, 99, 100, 105, 120, 122, 150

baby universe, *see* universe, baby
beautiful theory, *see* theory, beautiful
beauty, 2, 3, 5, 19, 26–27, 32–35, 52, 121, 139, 143
becoming, 99, 111–113, 131, 154
beginning of the universe, *see* universe, beginning of the
beginning of time, *see* time, beginning of
behavior, xii, 20, 39, 47–48, 54, 61–62, 64–65, 67, 70, 73, 79, 86, 88–93, 96, 135, 146, 151

behavior, law of, *see* law of behavior
being, 33, 37, 63–65, 99, 111–113, 131, 148, 153
being, human, *see* Homo sapiens
belief, 2–4, 9, 36–37, 40, 47, 65, 121–124, 126–127, 129, 144
bible, 3, 4, 36
big bang, 1, 49, 50, 132–133, 144, 146
big bang scheme, *see* scheme, big bang
big crunch (*see also* universe, collapse of the), 49, 132–133, 144
biochemistry, 22, 107
biology, 22, 70, 83, 101, 107, 110, 130
birth of the universe, *see* universe, beginning of the
brain, 37–38, 40, 69
branch universe, *see* universe, branch

causation, 24, 31, 33–35, 104, 107–108
cause, 24, 29, 31, 66, 75, 104, 108, 110, 114, 139, 140, 149, 153–154
capricious cosmos, *see* cosmos, capricious

cell, 38, 40, 68, 70–73
chemistry, 70, 93, 102
circularity, 108, 136, 140
clairvoyance, 6, 46, 94
coexistence, 49, 139, 148
collapse of the universe, *see* universe, collapse of the
color, 2, 3, 55, 112–113, 122
comet, 21–22, 30
common sense, xi, 120, 127, 129
complex system, *see* system, complex
complexity, 7, 25, 38, 40–41, 66–69, 72, 74, 77, 110–111, 135
component (*see also* part), 6, 7, 26, 38–39, 57, 66–68, 110
comprehension, x, xi, 2–4, 47, 51, 69, 77, 89, 121–122, 126, 145
comprehension, mode of, x, xi, 2–4, 13, 20, 46, 122, 126–127, 129
conceivability, 5, 6, 13, 16, 23, 33–34, 46, 48–49, 66, 73, 127, 138, 148
concept, x, xi, xiii, 14, 25, 30, 34–35, 37, 39–41, 45, 49, 75, 77, 80–81, 88–89, 95, 105, 108, 112, 121, 123–124, 127, 129, 132–134, 136–137, 139, 140, 148, 153

conception, ix, xii, 22, 37, 100–101, 114, 153
conceptual framework, *see* framework, conceptual
condition (*see also* state), 8, 30, 78, 87–88, 91, 103
connection, 14, 51, 53, 57, 76, 132–133, 144, 147
consciousness, 2, 37–38, 40, 42, 69, 100, 122, 149, 155
conservatism, xii, 52, 63, 134, 140, 144, 153
consideration, metaphysical, *see* metaphysical consideration
consideration, scientific, *see* scientific consideration
consistency, 52, 55, 87, 128, 137
context, invariant, 105–107, 114
contracting universe, *see* universe, contracting
contradiction, xi,, xiii, 21, 35–37, 52, 62, 85–86, 90, 94, 128, 137
conventionality, 94–95, 102, 105–108, 110–111, 114, 139
correlation, quantum, *see* quantum correlation
cosmic oscillation scheme, *see* scheme, cosmic oscillation

159

Index

cosmic scale, *see* scale, cosmic
cosmogony, 53
cosmological scheme, *see* scheme, cosmological
cosmology, xii, 51–52, 55, 57–59, 144–145
cosmos (*see also* universe; universe as a whole), 49–54, 57, 92, 131–132, 144, 152
cosmos, capricious, x, xi, 51, 57, 61, 85–86, 90, 94–95, 109–111, 114
counterintuitiveness, 29, 71, 125
creationism, 23, 36, 101
creator, 37
criterion, 19, 20, 23, 25, 28, 35, 41, 52, 69, 101, 107

Darwin, Charles Robert, 101, 107
data, 5, 6, 10–16, 29, 38, 51–56, 58, 86, 102, 118, 122–123, 147, 150, 152
death of the universe, *see* universe, end of the
deity, 37, 128
density, 50, 131
description, 3, 4, 13, 15, 19, 24, 31, 36, 42, 49, 52, 57, 71, 76, 79, 109, 122, 124, 135, 138, 147, 152, 154

description, literal, 124–127, 129, 151
detection, 63, 93
determinacy, 49, 124, 135, 151
development, 39, 40, 42, 64–65, 78, 102, 125–126
dimension, 23, 99, 111–113, 145, 153–154
Dirac, Paul Adrien Maurice, 27–28
direct knowledge, *see* knowledge, direct
direction, 7–9, 29, 38, 50, 68, 79, 87, 118, 135, 149
disconnection, 136–139, 143
discovery, 27, 47, 62–64, 78–80, 93–94, 102, 110, 118, 121, 147
distance, 10–12, 14, 54, 76
domain of metaphysics, *see* metaphysics, domain of
domain of science, *see* science, domain of
domain of validity, *see* validity, domain of
Doppler, Christian Johann, 55
Doppler effect, *see* effect, Doppler

Earth, 1, 7, 9, 14, 22, 25, 30, 36, 64, 74–75, 85, 113
economy, xii, 41, 121, 124, 127, 129, 133–134, 136–137, 139, 140, 153

Index

Eddington, Arthur Stanley, 109
effect, Doppler, 55
effect, red shift, 54–55
effect (phenomenon), 2, 46–47, 53, 71, 111, 125, 151
effect (result), 24, 71–72, 139–140, 153
Einstein, Albert, 25, 75, 89, 153
electricity, 27, 38, 70, 75–76
elementary particle, *see* particle, elementary
emotion, 6, 37–38, 155
end of the universe, *see* universe, end of the
end of time, *see* time, end of
ensemble, universe, *see* universe ensemble
environment, 40, 65, 96, 101, 110, 114, 125
ESP, *see* perception, extrasensory
esthetics, 4, 26, 143
evidence, 9, 34, 95
evolution (*see also* process), 1, 36, 54, 64–65, 77–81, 89, 101, 107, 110, 123, 145–147
evolution, adaptive, 83, 114, 124, 130
evolution, law of, *see* law of evolution
evolution of the universe, *see* universe, evolution of the

example, archetypical, 13, 29
existence, xi, 5, 10, 13, 22, 24–25, 27, 34, 37, 41, 45–48, 56, 62–63, 76–77, 85–86, 88, 90, 94–95, 100–111, 113–114, 118, 120–121, 131, 138–140, 143, 147, 150, 152–155
existence, mode of, 34, 136–137, 141
expanding universe, *see* universe, expanding
experience, 10, 47, 112–113
experiment, 4, 6–13, 21, 26–27, 29, 35, 46–48, 54, 56, 75, 80, 86, 88, 104, 120, 123, 151–152
experimental input, 10
experimental result, 6–10, 13, 16, 28, 46, 53, 80, 119–120, 147, 150
explanation, xi, xii, 3–5, 19, 20, 22–25, 28–31, 45, 47, 51–52, 57, 61, 70, 75, 88, 94–95, 99–111, 113–114, 139, 143, 145, 152–154
explanation, anthropic, 95, 102–108, 110–114, 139, 143
explanation, mode of, 13
explanatory principle, *see* principle, explanatory
explosion, 9, 49, 50, 56, 131, 144

161

Index

extended Mach principle, *see* principle, extended Mach
extra large scale, *see* scale, extra large
extranaturality, 39, 40–42
extrasensory perception (ESP), *see* perception, extrasensory

falsity, 21, 27–28, 30, 145
falsifiability, 21–22, 28, 31, 56, 145
falsifiable theory, *see* theory, falsifiable
falsification, 53, 56, 137
feeling, 2–4, 6, 20, 26, 28, 31, 34, 37–38, 40, 42, 57, 101, 104, 122, 126, 135, 143, 149, 154–155
final state, *see* state, final
finiteness, 9, 51
fluctuation, quantum, *see* quantum fluctuation
foam, space-time, *see* space-time foam
force (*see also* influence; interaction; *and specifically* electricity; gravitation; magnetism), 25, 29, 30–31, 73–76, 79, 83, 85–86, 89, 102–104, 106, 146, 149
foundation, 3, 35, 59, 62, 80
framework, conceptual, xii, 35, 41, 137, 139–140, 153, 155
framework, organizing, 35, 53, 58, 75, 88
framework, philosophical, 33, 35, 41, 148
framework, unifying, 23, 57
framework of science, *see* science, framework of
fundamentality, ix, xii, 21–24, 30–31, 33–35, 48–49, 89, 94–95, 101–102, 104–111, 114, 118, 126, 135, 138, 143, 150–151, 154
future, 29, 51–58, 77, 92, 135, 138, 144

galaxy, 1, 48, 53–56, 65, 74, 85, 87, 93, 121
gas, 38–39, 91
gene, 36
generality, 21–22, 24–26, 28, 30–31, 75, 89, 101, 104, 107–108, 114, 154
genetics, 22
geology, 47
God, x, 3, 23, 34, 37, 42, 128
gravitation, 25, 29–31, 73, 75–76, 79, 85, 102–104, 106, 149
guiding principle, *see* principle, guiding

heredity, 22
history, 2, 16, 108, 145
holism, 51, 67–69, 72, 75–77, 81–82, 88, 104, 108, 145, 152
Homo sapiens (*also* human *and* humanity), ix–xii, 3, 5, 6, 20, 22, 27, 37–38, 65–67, 69, 88–89, 91–92, 95, 99–102, 105–111, 113–114, 117–118, 121, 126, 128, 143, 145, 149–150, 152–153
human, *see* Homo sapiens
human scale, *see* scale, human
humanity, *see* homo sapiens
hypothesis, 4, 12, 14, 70, 89, 106, 119

idea, 6, 23, 26, 49, 54, 71, 103, 128, 135, 137, 143, 148
idealism, 63–64, 66, 68, 81–82, 88, 100, 118, 146, 152
implication, logical, 20–21, 24, 30–31, 103–104, 106–108, 110–111, 113, 154
implosion, 49, 132, 144
inconsistency, 35, 40
indeterminacy, quantum, *see* quantum indeterminacy
indirect knowledge, *see* knowledge, indirect
inertia, 73–76, 86–89, 95, 146, 148
infinity, 134, 136
inflationary scheme, *see* scheme, inflationary
influence, physical (*see also* force; interaction), 47, 49, 50, 73, 75–76, 88
information, 51, 69, 122
initial state, *see* state, initial
input, experimental, *see* experimental input
insight, 10, 57–58, 65, 127, 129
inspiration, 10, 122–123
intelligence, 2, 62–65
interaction (*see also* force; influence; *and specifically* electricity; gravitation; magnetism), 5, 6, 13, 16, 23, 33–34, 46, 48–50, 53, 56, 66, 69, 73, 75, 79, 100, 110, 122, 127, 135, 146, 148, 151
interpretation, 35, 49, 53–55, 58–59, 63–64, 102, 109, 139–140, 142, 146, 148, 153, 155
interrelation, 51, 53, 57, 66–68, 71, 104, 144
interval, time, *see* time interval
intuition, 2, 10, 122–123, 125–126

163

Index

investigation, 6, 9, 10, 13, 33–34, 40, 45–46, 48, 53, 56, 67–68, 70, 73, 76–78, 86, 90, 92, 108–109, 119, 121, 123, 128
investigator, 6, 13, 56, 108–109, 152
irrationality, 20, 22, 26, 123, 127
irreproducibility, 6, 13, 34, 45–48, 50–52, 57, 61, 94, 110–111, 153, 155
island universe, *see* universe, island
isolability, 76–77
isolated system, *see* system, isolated
isolation, 67, 73, 75–77, 81, 89, 151

Kepler, Johannes, 13, 15, 146–147
Kepler's laws, *see* laws, Kepler's
knowledge, x, 1, 2, 4, 19, 28, 46, 51–52, 55–56, 65, 68, 75–77, 86–89, 92, 100, 102, 104, 107, 109–110, 119, 121–123, 125, 127, 129, 138, 145, 148
knowledge, direct, 122–123
knowledge, indirect, 122–123, 127, 129

large scale, *see* scale, large
law, x, xi, 1, 9, 10, 12–13, 15–16, 19–22, 24, 26, 29–31, 45–47, 51–53, 55, 57–58, 62, 64–65, 77–81, 85–86, 88, 91–93, 95, 118, 136, 147, 149, 151
law of behavior, 53, 89, 134–136
law of evolution, 77, 79, 80–81, 146–147
law of nature, 13, 15–17, 19, 21, 29, 31, 34, 42, 51–52, 61–66, 79, 81, 85–90, 94–96, 106–108, 110–111, 118, 134–135, 137, 140, 144–147, 151–154
lawfulness, xii, 61–62, 79, 87, 90, 92–93, 95
lawlessness, x, xi, 46, 51, 57, 61, 90–94, 134
laws, Kepler's, 13, 15–16, 21–22, 24–25, 29–31, 49, 70, 146, 149
laws, Newton's (*see also* Newton, Isaac), 29–31, 62, 73–75, 79, 85–87, 146, 149
life, 1–3, 39, 40, 42–43, 65–68, 92, 132, 134
light, 3, 51, 54–55, 64, 93
literal description, *see* description, literal

location (*see also* position), 7–9, 14, 38, 87, 93, 123, 125–126, 136, 147
logic, xi, 9, 20–21, 24, 30–31, 35, 40, 52, 90, 103–107, 110, 113, 120, 154
love, 2, 5

Mach, Ernst, 75–76
Mach principle, *see* principle, Mach
Mach principle, extended, *see* principle, extended Mach
macroscopic scale, *see* scale, macroscopic
magnetism, 75–76
many worlds, 49, 59, 139–140, 142, 148, 151, 153
mass, 25, 27, 29, 30, 74, 147, 149
materiality, x, 2, 5, 6, 8, 34, 38, 41, 45, 51, 63, 66, 112, 119–121, 133–134, 136–137, 139, 146, 148–149, 151
material universe, *see* universe, material
mathematics, 24, 30, 49, 151
matter, 1, 29, 33, 42, 54, 74–76, 95, 148
measure, 5, 12, 29, 74, 112–113, 118, 122, 145, 148
measurement, 5, 100, 118–119, 137, 150

mechanics, 30, 149
medicine, 65
metaphysical consideration, x, xi, 34–35, 43, 89, 117, 133
metaphysical position (*see also* world view *and specifically* conservatism; creationism; holism; idealism; positivism; rationalism; realism; reductionism; solipsism), xii, xiii, 35–36, 62–65, 67–69, 72, 77, 81, 88, 100, 118, 120, 128, 145–146, 148, 150, 152
metaphysics, x, xii, 33–36, 41–42, 48–53, 57, 59, 60, 62, 75, 88–89, 108–109, 118, 128, 131–134, 136, 143, 145, 148
metaphysics, domain of, x, 34–35, 41, 89, 95
metatime, 50, 132–134, 140, 148
microscopic scale, *see* scale, microscopic
mind, x, 2, 6, 33, 37–38, 40, 42, 62–65, 81, 99, 100, 117–118, 122, 126, 146, 148, 155
mode of comprehension, *see* comprehension, mode of
mode of existence, *see* existence, mode of

mode of explanation, *see* explanation, mode of
mode of understanding, *see* understanding, mode of
model, xi, 16, 108
molecule, 22, 38–40, 49, 68, 70, 72, 91, 93, 107, 151
moon, 15, 29, 48, 64, 67, 78–79, 85–86, 147
morals, 2
motion, 7, 9, 13–15, 21–22, 24–25, 27, 29–31, 38, 62, 70, 73–74, 79, 86, 88, 146–147, 149
motivation, 121, 123, 127, 129
myth, 3, 4

naturality, 23–24
nature, ix–xii, 1, 5, 6, 8, 9, 13, 16, 19, 23–24, 26, 31, 33–41, 45–50, 54, 62, 66–73, 76–78, 80–81, 83, 92–95, 99–101, 105–109, 114, 118–123, 126–129, 133–138, 144–148, 150–153, 155
nature, aspect of, ix, xi, 4, 6, 9, 10, 12–13, 16, 19, 23, 25, 31, 33, 40, 45, 47, 62, 66–67, 69, 70, 72, 87, 93, 99, 101–102, 104–109, 111, 114, 121, 123–125, 128, 143, 151–154

nature, law of, *see* law of nature
nature, phenomenon of, ix, xi, 6, 9, 13, 16, 19, 21, 28, 45, 47, 66–67, 72, 93, 101–102, 105, 107, 109, 114, 119, 143, 145, 147, 149–152, 155
nature, reduction of, 67, 69, 81
Newton, Isaac, 29, 30, 49, 63, 70, 75, 87, 149
Newton's laws, *see* laws, Newton's
nonexistence, 29, 56, 65, 103, 110
nonisolability, 73
nonmateriality, 2
nonphysicality, 121–122
nonscience, x, xi, 2–4, 13, 20, 46, 122, 126–127, 129
nonseparability, 71–72, 76–77, 81, 106
nonseparability, quantum, *see* quantum nonseparability
nontranscendence, xiii, 37, 39–42, 127–128, 155
nucleus, 27, 70, 76, 91, 93, 103, 119–120

objectification, 25
objectivity, xii, 3, 6, 13, 20, 24, 26, 40, 64–65, 81,

Index

104–105, 107, 114, 117–119, 121, 128, 144, 152
objective reality, *see* reality, objective
observation, 4, 5, 27, 34–35, 48, 54, 56, 69–73, 76, 86–87, 92–93, 100, 102, 118–119, 122–123, 125, 146, 148, 150
observed, 34, 53, 69–72, 77, 81–82, 88, 103, 149–150
observer, 14, 34, 62–63, 69–72, 77, 81–82, 88, 102, 118, 121, 124–125, 128, 130, 149–152
orbit, 14–15, 25, 147
order, x, 1, 9, 10, 13–16, 30, 34, 45–47, 51–52, 57, 61, 64–65, 77–78, 80–81, 85–86, 88, 90–96, 100, 109–110, 114, 118–119, 147, 150–151
orderlessness, x, xi, 46, 51, 57, 61, 86, 90–96, 110–111
ordinary scale, *see* scale, ordinary
organism, 22, 39, 40, 67
organizing framework, *see* framework, organizing
origin, xii, 75, 86–89, 95, 109, 145, 148
origin of the universe, *see* universe, origin of the

paradox, xi, 90, 94, 105
parapsychological phenomenon, 6, 10, 45–46, 94, 111
part (*see also* component), 4, 5, 21, 28, 34, 37–38, 40, 61, 67–69, 73, 76–77, 80–81, 88, 90, 92, 96, 101, 104, 108–109, 122, 125–126, 132, 134, 146, 148, 151–152, 155
partially hidden reality, *see* reality, partially hidden
particle, elementary, 8, 23, 27–28, 53, 56, 70–71, 91, 119–120, 125, 135, 145
past, 51–58, 135, 138, 144
perceived reality, *see* reality, perceived
perception, xi, 24–25, 31, 33–35, 49, 63–64, 74, 100, 104–105, 114, 123–124, 126, 137, 146, 154–155
perception, extrasensory (ESP), 6, 34–35, 45, 94
phenomenon, 3, 5, 22, 27, 30, 34–36, 45–46, 52–53, 64, 67, 69–73, 81, 94, 111, 118–119, 124–125, 149
phenomenon, parapsychological *see* parapsychological phenomenon

167

Index

phenomenon, unique, 7, 48, 50, 52, 57, 61, 152–153, 155
phenomenon of nature, *see* nature, phenomenon of
phenomenon of reality, *see* reality, phenomenon of
phenomenon of the universe, *see* universe, phenomenon of the
philosophical framework, *see* framework, philosophical
philosophy (*see also specifically* metaphysics), ix, xii, xiii, 33, 36, 59, 89, 111–112, 117, 148
physicality, 36, 40, 74, 110, 118, 121–123, 150
physicist, ix, xi, 4, 16, 25, 40, 42, 48, 57, 59, 60, 129
physics, ix, xi, 16, 32, 42, 107
Planck, Max, 131, 135
Planck scale, *see* scale, Planck
planet, 1, 13–15, 21–22, 24, 29, 30, 48, 62, 70–71, 78–79, 85–86, 101–102, 128, 146–147, 149
position (*see also* location), xi, 25, 78–79, 125
position, metaphysical, *see* metaphysical position
positivism, 118–120, 150, 152
possibility, quantum, *see* quantum possibility
predictability, 4, 6, 9, 10, 12–13, 15, 19, 33–34, 45–47, 51, 57, 61, 66, 69, 77–78, 80, 85, 92–94, 95, 99, 100, 109–110, 147, 151, 153
prediction, 9, 10, 12–13, 15–16, 19, 21–22, 27–29, 31, 45, 53, 55–56, 58, 79, 80, 91, 93, 120, 145, 147, 151
predictivity, 19, 53, 56–58
present, 5, 51, 53–54, 56–58, 73, 102, 112, 117, 126, 138, 144
pressure, 7, 39, 50, 91, 121, 131
principle, 5, 19, 41, 75, 108–109, 148
principle, anthropic, xi, xii, 22, 95, 101–111, 113–115, 143
principle, explanatory, xi
principle, guiding, 75, 88–89
principle, Mach, 75–76, 83, 86–89, 95–96, 146, 148
principle, extended Mach, xii, 87–90, 94–95, 145
probability, 49, 124, 135, 151
process (*see also* evolution), 36, 77–78, 80, 145
proof, 9, 34, 110
property (trait), 14–15, 20–

26, 28–29, 31, 38–40, 42, 53–54, 56, 63, 68, 73–76, 81, 91, 108, 125, 144–147, 154
public, ix, 3, 112

quantum alternative, xii, 139–140, 153
quantum aspect, 111, 124–125, 128, 130, 135–136, 141, 150
quantum correlation, 76, 81, 111, 135, 151
quantum fluctuation, 135, 139–141
quantum indeterminacy, 138
quantum nonseparability, 71–72, 76–77, 81–82, 135, 151
quantum possibility, 49, 124, 138–139, 148, 151
quantum uncertainty, 138
quantum unpredictability, 91, 93, 95–96, 110–111, 135, 138, 151
quantum theory, *see* theory, quantum
quasi-isolated system, *see* system, quasi-isolated

rationality, 1, 26
rationalism, 2
realism, 62–64, 66, 68, 81–82, 88, 100, 118–121, 146, 150, 152

reality, ix, xi, xii, 33, 35, 37, 41, 49, 55, 62–63, 67, 114, 117–122, 124–125, 127–129, 138–139, 148–149, 152, 155
reality, aspect of, 123–124, 126–127, 129–130
reality, objective, xi, xiii, 118–119, 121–129, 146, 149–150
reality, partially hidden, 126–129, 149
reality, perceived, 123–128, 149–151
reality, phenomenon of, 117, 127
reason, 1, 5, 13, 19, 20, 31, 40, 47–48, 54–55, 63, 67, 80, 87, 89, 102–103, 105, 108, 110, 113, 117, 124, 126, 132, 140, 144, 154
reasonableness, 58, 63, 65, 87, 100, 120–121, 123–124, 127, 129, 144
red shift effect, *see* effect, red shift
reduction of nature, *see* nature, reduction of
reductionism, 67–69, 72, 75–77, 80–82, 88, 145, 150, 152–153
reference (relation), 53, 75, 86, 88–89
rejection of theory, *see* theory, rejection of

Index

relation, 10, 12, 15, 35, 40, 66, 91, 126, 132–134, 137, 139, 150
relativity, theory of, *see* theory of relativity
religion, 2, 3, 35, 37, 42, 108, 155
reproducibility, 4, 6–9, 12–13, 16, 19, 33–34, 45–48, 51–52, 54, 61, 66, 69, 77–80, 85–86, 99, 100, 121, 152–153
research, 5, 53, 70, 148
rest, 10, 12, 29, 74, 146, 149
result, experimental, *see* experimental result
retrodiction, 53–55, 79
rotation, 7, 8

satisfaction, 4, 5, 19, 20, 24, 31, 40, 66, 68, 94, 99–101, 103, 107, 111, 114, 119, 153
scale, astronomical, 53, 70, 93, 124
scale, atomic, *see* atom
scale, cosmic, 53, 93
scale, extra large, 53, 124
scale, human, 47, 93, 94, 110
scale, large, 23, 72, 80, 111, 124
scale, macroscopic, 22, 39
scale, microscopic, 22, 70
scale, molecular, *see* molecule
scale, nuclear, *see* nucleus
scale, ordinary, 70–71, 80, 125
scale, Planck, 135–139, 143, 150
scale, small, 70–71, 80–81, 93, 110, 135
scale, subatomic, 49, 145, 151
scale, submicroscopic, 53, 70–71, 76, 93, 95, 110–111, 124, 135, 143, 151
scale, subnuclear, 91, 135, 145
scheme, 3, 4, 144, 152, 154
scheme, big bang, 50, 53, 55, 57, 131–134, 140, 144, 153
scheme, cosmic oscillation, 49, 132–134
scheme, cosmological, 3, 4, 52–58, 102, 131–132, 134, 136–139, 140–141, 144, 146, 152–153
scheme, inflationary, 50, 53, 57, 146
scheme, self-generating universe, xii, 136–140, 153
Schrödinger, Erwin, 27
science, ix–xiii, 1–6, 9, 10, 13, 16, 19, 20, 22–24, 26–27, 32–37, 40–42, 45–52, 57–61, 63, 65–66, 68–71, 75–77, 80–81, 88–89, 93–95, 99–102, 104–

Index

105, 108–109, 114, 117–123, 126–129, 133–134, 137–139, 144–145, 148–155
science, domain of, x, xii, 2, 3, 5, 6, 10, 33–34, 41, 45, 118, 126, 136, 151–154
science, framework of, x, 46, 50–51, 57, 93, 101–102, 105, 109, 114, 131, 134, 143
scientific consideration, 34, 41
scientist, ix, xii, xiii, 2, 4, 22–26, 30–31, 34–35, 37, 41, 49, 62–63, 76–77, 93, 104, 109, 118–119, 122–123, 128, 143–144
self-consistency, 35, 52, 128, 137–138
self-contradiction, xiii, 36
self-generating universe scheme, *see* scheme, self-generating universe
self-reference, 108, 136, 140
sense (sensation), 63–65, 69, 100, 105, 117–118, 120–123, 146, 150
sense, common, *see* common sense
separability, 68, 75, 88, 152–153
separation, 30, 68–73, 76–77, 80–81, 149–151

simplicity, 13, 17, 25–26, 30–31, 33–35, 40, 52, 66–69, 76, 92, 99, 104, 107, 113–114, 128, 139, 144, 147, 150, 154
simplification, 4, 72, 80
small scale, *see* scale, small
solar system, *see* system, solar
solipsism, 120
soul, 1, 99
source (origin), 2, 4, 29, 55, 64–65, 99, 119, 149
space, 7, 14, 23, 25, 56, 71, 77, 87, 89, 92–93, 99, 106, 110–114, 126, 135, 147, 150–151, 153
space-time, 131, 135–136, 139–140, 143, 153
space-time foam, 135, 141, 143
spectrum, 3
speed, 3, 14–15, 25, 29, 38, 50–51, 54, 79
star, 1, 23, 48, 54–55, 64, 85–86, 102, 121
state (*see also* condition), 7, 9, 38, 51, 53, 56–57, 74, 78–79, 146–147
state, final, 79, 133
state, initial, 77–81, 133, 146–147
subatomic scale, *see* scale, subatomic
subjectivity, 3, 22, 24–26, 64, 68, 104–106, 114, 122, 127–128

submicroscopic scale, *see* scale, submicroscopic
subnuclear scale, *see* scale, subnuclear
Sun, 7, 14–15, 24, 30, 56, 64, 78, 85, 147
surroundings, 47, 67, 69, 71–73, 76–77, 81, 87–89, 100, 146, 149–151, 154
survival, 36, 39, 40, 64–65, 102, 110, 123–124, 127, 129
symmetry, 1, 16, 32
synthesis, 67, 81, 152–153
system, 15, 27, 30, 40, 43, 47–49, 54, 73, 75–78, 80–81, 85–87, 89, 90, 92, 96, 100, 113, 122, 146–147, 151
system, complex, 37–40, 42
system, isolated, 47, 73, 75–77, 146, 151
system, quasi-isolated, xii, 77–81, 85–90, 92, 95, 145–147, 151, 154
system, solar, 7, 14–15, 24, 51, 56, 78–79, 85, 87

taste, 2, 25, 30, 33–36, 40–41, 123
technology, ix, 5, 29, 121, 148
telekinesis, 6, 46, 94
teleology, 37, 108
telepathy, 6, 46–47, 94
telescopy, 51, 54
temperature, 39, 50, 91, 121, 131
test, 21, 28, 56, 145
theology, 36, 42
theory, 4, 19, 21–24, 26–32, 35, 45–46, 48–49, 52–53, 57–58, 60, 63, 75, 88, 94, 101–102, 109, 122–123, 134, 137, 143–145, 149, 151–152, 154
theory, acceptability of, 20, 24, 101, 154
theory, acceptable, 20–26, 28, 31, 101
theory, acceptance of, 19
theory, beautiful, 26–27, 31, 143–144, 154
theory, falsifiable, 28–29, 31, 145, 154
theory, quantum, 48–49, 59, 71, 76, 80, 91, 93, 95, 107–108, 117, 124–127, 129, 135–136, 138–140, 142–144, 148, 151, 153
theory, rejection of, 19
theory, unacceptable, 21–23
theory, unfalsifiable, 28
Theory of Everything (TOE), x, 57, 60, 154
theory of relativity, 25, 75, 89, 117, 125, 153
thought, 37–38, 40, 105, 117, 120, 122, 125, 155

Index

time, 7, 8, 14–15, 23, 25, 29, 38, 40, 50, 54–55, 64, 77–79, 86, 89, 92–93, 99, 101–102, 106, 110–111, 113–114, 119–120, 123, 126, 132–133, 135–136, 139–140, 145, 147–148, 150–151, 153–154
time, beginning of, 133
time, closed, 133–134
time, end of, 133
time interval, 9–12, 14, 92, 135, 137, 147, 150
TOE, *see* theory of everything
tone, 55, 112
transcendence, xiii, 37, 39, 40–42, 101, 127–129, 155
truth, 5, 8, 19, 21, 23, 28, 35–36, 40–41, 45, 92, 104, 145

unacceptable theory, *see* theory, unacceptable
uncertainty, quantum, *see* quantum uncertainty
unconventionality, xi, xii, 95
understanding, x, 1–5, 9, 13–14, 16, 19, 20, 26, 30–31, 33–34, 37, 39, 46–47, 51, 53, 62, 66–70, 73–75, 81, 89, 99–102, 105, 109, 112, 117, 119–120, 123–124, 126–129, 135–136, 144–145, 151–153
understanding, mode of, x, xi, 2, 46, 126–127, 129
undetectability, 23
unexplainability, x, xi, 46, 51–52, 57, 61, 95
unfalsifiability, 28
unfalsifiable theory, *see* theory, unfalsifiable
unification, 3, 13, 15, 19, 25–26, 30–31, 104, 107, 114, 144, 147, 154
unifying framework, *see* framework, unifying
unique phenomenon, *see* phenomenon, unique
unity, 67, 139
universality, 22, 25, 29, 30, 49, 73, 79, 101, 146, 149
universe (*see also* cosmos), ix, xi, xii, 5, 6, 23, 25, 30, 37, 45, 48–57, 61, 74–75, 81, 86, 88–90, 95–96, 100, 104, 107–108, 131–141, 143–144, 146, 148, 153
universe, age of the, 102
universe, aspect of the, x, 2, 22, 34, 37, 51, 53, 57, 61, 63, 92–95, 104, 106, 110, 132, 137, 140, 144
universe, baby, 136–140, 143, 151, 153

173

Index

universe, beginning of the (*also* birth *and* coming into being of the), xii, 3, 4, 23, 37, 49, 50, 55, 81, 131–134, 136–137, 139–140, 144, 148, 153
universe, branch, 49, 139, 148
universe, collapse of the (*see also* big crunch), 132–133, 144
universe, contracting, 49, 132, 144
universe, end of the (*also* death of the), 3, 49, 132, 134, 144, 148
universe, evolution of the, 3, 23, 49, 53–56, 58, 93, 132–133, 139
universe, expanding, 1, 49, 50, 55, 131–133, 144, 146
universe, island, 34, 50, 146
universe, material, x, 2–6, 13, 16, 33–34, 37, 46, 48, 66, 73, 101, 127, 148
universe, origin of the, 53, 55, 57–58, 137
universe, phenomenon of the, x, 2, 51, 53, 57, 61, 104, 137, 145
universe, self-generating, *see* scheme, self-generating universe
universe as a whole (*see also* cosmos), x–xii, 2–4, 22, 48, 50–53, 57–58, 61, 80–81, 86–90, 92, 94–95, 104, 144–145, 152, 154–155
universe ensemble, 48, 50, 60, 131, 134, 141
unpredictability, 10, 13, 34, 46, 51, 57, 61, 91–94, 110, 119, 151
unpredictability, quantum, *see* quantum unpredictability
utility, 40, 72, 77, 80, 102, 121, 123, 127, 129

vacuum, 135, 141, 151
validity, 3, 4, 6, 15, 21–22, 24, 28–30, 40–41, 52, 56, 62–64, 68, 71, 74, 87, 89, 92, 94–95, 100–101, 103, 105, 109, 114, 124–126, 137, 140, 144, 147, 152–153
validity, domain of, 2, 3
velocity, 7, 8, 29, 38, 74, 79
view, world, *see* world view
volume, 39, 87, 91, 110

wave, 3, 51
weight, 29
whole, universe as a, *see* universe as a whole

wholeness, 66–68, 81, 145
world, ix, xi, 1, 3, 11, 19, 20, 33, 45, 51, 62–64, 85, 118, 121–122, 131, 136–137, 146, 148
world view, xiii, 22, 24–25, 30, 33–35, 37, 40–41, 62–65, 67, 81, 95, 101, 108, 118, 127–129, 155
worlds, many, *see* many worlds

ABOUT THE AUTHOR

Joe Rosen received his Ph.D. in theoretical physics from the Hebrew University of Jerusalem in 1967. Although he set out in the field of elementary-particle physics, he has done research in other fields as well, eventually gravitating to his present main interest in the foundations of physics. The dominant theme in his physics research has been symmetry as a unifying principle. Joe Rosen has written and edited six books on physics and symmetry and has published a large number of articles in professional journals. Music is his major hobby: playing clarinet, composing and arranging pieces for wind ensemble and for chamber groups, and conducting.

A permanent member of the faculty of the School of Physics and Astronomy of Tel Aviv University, Joe Rosen has held positions at Boston University, Brown University, the University of North Carolina at Chapel Hill, and the Catholic University of America.